法国旧藏中国家具实例
CHINESE FURNITURE
A Series of Examples from Collections in France

中国家具经典图书辑丛

法国旧藏中国家具实例

赫伯特·塞斯辛基

故宫出版社

CHINESE FURNITURE

A SERIES OF EXAMPLES FROM
COLLECTIONS IN FRANCE
WITH AN INTRODUCTION BY

HERBERT CESCINSKY

With Fifty-four Collotype
Plates and Ten Half-tones

LONDON: BENN BROTHERS, LIMITED
8, BOUVERIE STREET, E.C.4
1922

出 版 前 言

本书根据1922年英文版《中国家具》翻译而成。

赫伯特·塞斯辛基是活跃于20世纪初的英国学者，谙熟西方家具、室内装饰的风格和历史，为本书撰写的前言显示了他精深渊博的学识，我们很惊诧于这位近百年前的学者，竟然对中国家具乃至中国艺术或东方艺术，有如此深刻的认识，因为很多观点在今天仍然适用，光辉熠熠。当然，作者的个别观点，略有局限，但瑕不掩瑜。

书中收录家具来源，主要为奥迪隆·罗什先生、沃尔希先生、查尔斯·维涅先生、L.万尼克先生、保罗·马隆先生、圣约翰·奥德利先生以及中国通运公司（张静江先生开设）、卢吴公司（卢芹斋先生等开设）的旧藏。家具以清代早期尤其是康熙时期所制为特色，它们多是当时作为贸易品出口，此外也有一些珍贵明代或清中晚期制品。历经沧桑，这些家具除了部分保存在博物馆外，大多已散佚无存，这些图版尤显珍贵。

本书尽量反映原书风貌，略作的改动说明如下：

1.书名改为《法国旧藏中国家具实例》，因书中所录家具皆来自法国旧藏，同时区别于另一本《中国家具》（1926年，德文版，莫里斯·杜邦著，该书更名为《欧洲旧藏中国家具实例》，于2013年由故宫出版社出版）。

2.增加中文翻译。原英文引言的插图加入中文引言中。

3.添加中英文出版前言及目录。

4.在文字页添加页码。

5.原书图题印在图版对开页上，此次出版移前一页，原有随图版横置的图题也统一转正。

FOREWORD

This book is the Chinese version of Chinese Furniture which was published in 1922.

As a famous British scholar Herbert Cescinsky, who was active in early twentieth century, was familiar with the style and history of Western furniture and interior decoration. The introduction of the book written by Herbert Cescinsky completely shows his profound knowledge. We were astonished at the author's deep understanding of Chinese furniture, even of the Chinese art or Oriental art. Most of the viewpoints still hold true at present and by no means outdated. Admittedly, some individual viewpoints of the author were somewhat limited, but they don't outweigh the merits.

The origin of the furniture in the book included the collections of O. Roche, Worch, Charles Vignier, L. Wannieck, Paul Mallon, St. John Audley, the Chinese Tonying Co. which was set up by Chang Ching-chiang, and Messrs. Loo & Co. which C.T.Loo was a mainly founder. As the traded commodity, the majority of the furniture belongs to the early Qing dynasty, even the period of Kangxi's Reign. Furthermore, some valuable furniture of Ming or Late Qing dynasty was also contained. With some exceptions preserved in the museums, others were mostly scattered and lost due to vicissitudes, therefore the plates become increasingly precious for this reason.

Based on the principle of keeping original look, the slightly changes were explained below:

1. The Chinese name of the book is translated into *Chinese Furniture: a Series of Examples from Collections in France*, and is distinguished from another German book of Chinese Furniture written by Maurice Dupont published in 1926. The book will also be published by the Forbidden City Publishing House in 2013, and the name will changes to *Chinese Furniture: a Series of Examples from Collections in Europe*.

2. The Chinese translation will be added in the book, and the original figures of introduction in English will be moved to Chinese version.

3. The foreword and contents, both in Chinese and English, are new for the book.

4. The page numbers will be added to the furniture introduction.

5. The page of furniture introduction was moved one page ahead, other than a folio before, and the thwartwise text will be turned around for reading easily.

目　录

出版前言 7

出版说明 11

引　言 13

图　版 33

CONTENTS

Foreword 8

Publishers'Note 12

Introduction 25

Plates 33

出 版 说 明

中国艺术，尤其那些具有更为简练、质朴风貌的领域，它们的关注度越来越高，已是当前的一种趋势。家具可能是近来极受鉴赏家及藏家们瞩目的中国艺术门类。但很遗憾，关于这个门类，目前并没有出版任何书籍，相关的专家也比较少。这一领域需要长期细致入微的调研与考证，我们期望这些成果能够及时地发布。此刻，借助出版法国商人和鉴赏家的这部分收藏品的影像资料，我们将弥补这一空白。这批资料出现得适逢其时，更不必提那些异常珍贵、甚至独一无二的实例，它们充满美感，引人关注，收藏者们无不梦想着能够拥有。

乾隆宝座的影像承蒙维多利亚及阿尔伯特博物馆的惠准得以复制。

PUBLISHERS' NOTE

The present time is remarkable for a growing interest in Chinese art, and particularly in the more simple and austere aspects of it. Furniture is perhaps the department of Chinese art which has most recently attracted the attention of connoisseurs and collectors. But unfortunately no books have been published on the subject, and there is available only a very small stock of expert information. An exhaustive work would require some years of investigation and research ; and it is our hope in due course to publish the results of such inquiry. Meanwhile, a gap may be filled by this series of reproductions of specimens in the hands of dealers and connoisseurs in France—specimens which happened to be available at the moment in a collected form, and which, while not claiming to include many of the rarer and indeed unique examples, all possess points of interest and beauty, and are such as the ordinary collector may hope to acquire.

The Ch'ien-lung throne is reproduced by kind permission of the Victoria and Albert Museum.

引 言

近年来，我们才了解了一些关于中国艺术方面的知识，当然，这只是凤毛麟角。我们拥有了瓷器、漆器、刺绣、木器、毛毯及其他产品，它们来自北京、南京、广东、福州、厦门，乃至欧洲本地，多是在约翰公司和荷兰东印度公司成立后，但有些，甚至早到在通往东方的好望角航线发现之前。在很久之前，东方艺术品就通过诸如威尼斯和热那亚等贸易城市来到西方，这一直持续到伊丽莎白女王统治时期。1600年，英国东印度公司获得特许，与东方的贸易日渐频繁且逐步扩大，但此时我们对于中国或日本艺术品还知之甚少，时常混淆（事实上在东方艺术研究中它们有着本质的区别）。以至于在1641年发行的《东印度贸易：关于东印度不同港口的真实描述》一书中，会提及日本出口"各式各样的漆制品"，同时强调"通常被带回欧洲的商品涉及柜、小床、橱，以及形形色色的家居用品"。

对于中国，古文献中极少涉及，只是偶见一些关于华夏[*]帝国的零星记载。日本的影响更大，以至于我们用"Japanning"或者"Jappan work"来命名漆器。类似的例子不胜枚举，比如在欧洲，某类器物的名称，通常使用这类器物的中转贸易站，而非原产地。因此，在早期"万丹工"被用来命名那些款彩漆器（原产地通常是中国或韩国），这是为了纪念荷兰商人在爪哇岛的贸易站万丹，此地距内陆9英里多，直到1817年因被西冷取代而遗弃。

事实上尽管这些器物已经出现在了这个国家，例如大屏风，以及习见的方柜，英国工匠们通常在下面安装描金雕花底座，人们对这些东方漆器作品依然知之甚少，但这并不妨碍某些江湖骗子自诩为"漆"或者"漆器"专家。"金球"（译者注："金球"概为伦敦圣詹姆斯市场某地名）的约翰·斯托克和牛津的乔治·派克在1688年出版了一本小册子，书中在介绍了所有艺术的秘密之后，于《给读者及从业人员的一封信》中强调了"贵族和上流人士通过拥有一整套'日本漆器'来彰显其身份（这些冒牌专家是如此的势利），与之相对，其他人只能不得不满足于一件屏风、一个化妆盒或海碗，又或者是无人知晓的老物件"。

本书正是要介绍这些所谓的"老物件"，但在详细解释书中的图版之前，笔者想简述一下中日两国的漆艺，特别是两国作品的不同之处，以期对读者有所裨益。

东方的漆是天然漆树的树脂（漆酶），为漆树科漆树属。从树上新渗出的漆是可溶的，但干固后即便

---

[*] 原文中"Cathay"一词，可能源于俄语"KHITÂÏ"（译者注："KHITÂÏ"可直译为"契丹"，在古代西方，通常用以表达"中国"之意，为避免歧义，本文译为"华夏"）。

是诸如酒精这种强烈的介质都无法影响到它。欧洲漆器，有号称仿制者，也只是简略地画一些装饰性图案，这尚算好的，更次者，不过是草率地涂鸦，用刷子或橡胶在表面使用虫胶漆弄出点光泽来。图1及图2所示平几是比较罕见的例子，毫无疑问它是欧洲漆器作品，且达到了较完美的水平，几乎可与中国或日本作品一较高下，但即使拥有如此质量，在东方的概念中，仍然称不上真正的漆器。

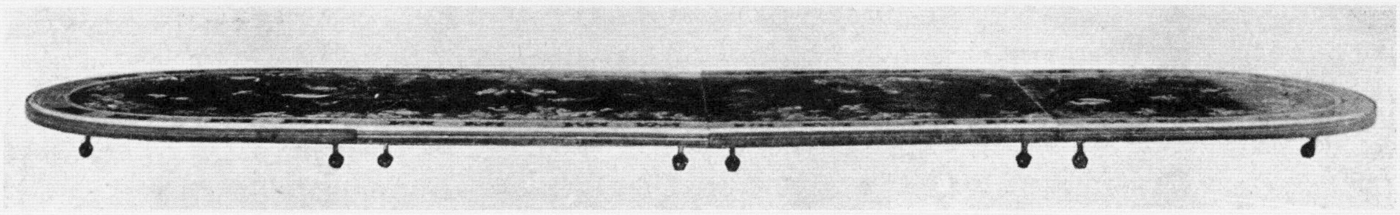

图1

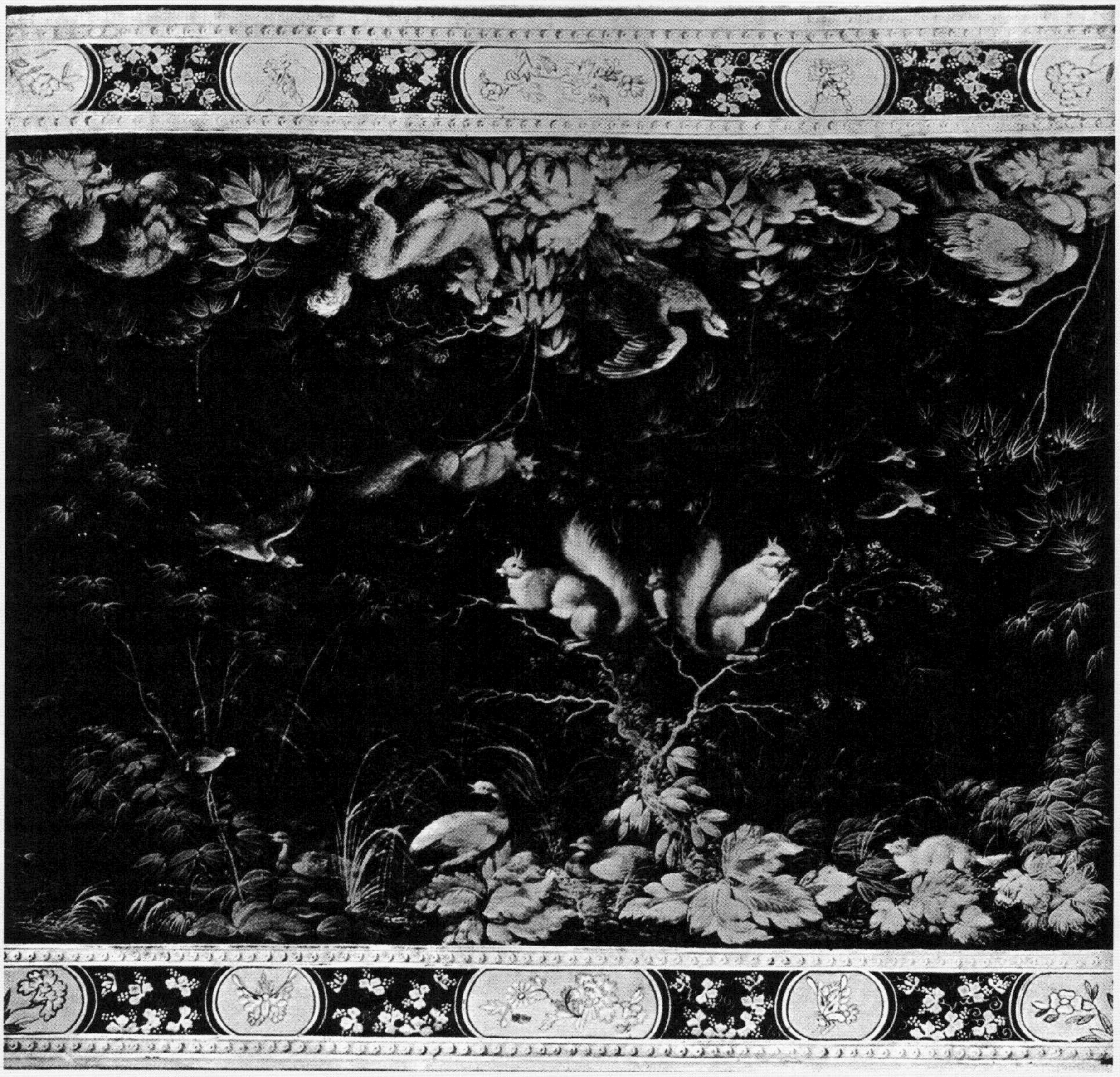

图2

这里来简要介绍一下中国和日本家具的一些主要区别，如图3这件立于英国制描金雕花底座上，被称为中国柜的作品，可作为一个典型。许多诸如此类叫做中国柜的作品，实际上来自日本。橱柜底座的支撑腿或为弯曲，或为直柱。底座使柜格可以立在地上，并保持离地3英尺。通常的处理方式是腿足上端以牙板固定，使得完成的底座能够保证橱柜的稳固性。

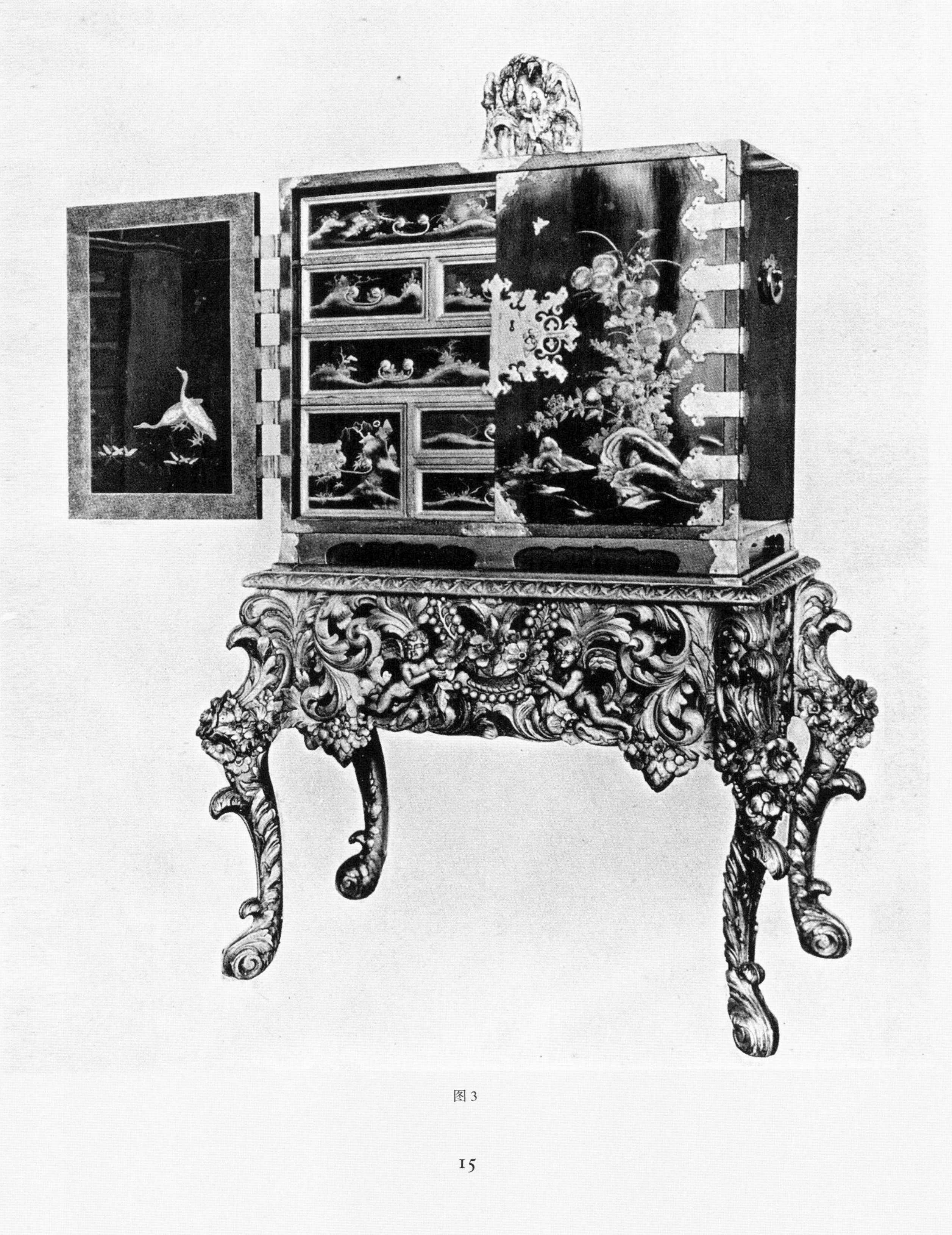

图3

这一点正好引出了中日家具间的第一个区别，中国人坐在椅或凳上，而日本人则席地而坐。对于后者而言，每件家具都依照比例降低高度。日本人的坐姿使得他们相对于中国官员或士绅，正好低了3英尺。

而中日漆器的另一个重要区别，只有在对两国产品有细致了解后才能够充分体会。显而易见，时代越早的漆器，越难以理解。正如早期英国橡木制品上玻璃光或蜡光的效果，通常称为"patine"（译者注：意为古木器上的光泽），这实际上是历经了几个世纪的使用后，时间赋予它们的柔和光泽与稳定外貌，本不是其固有的，正如同时期的中国漆器。这些对于现在即使能够熟练制作日本漆器的漆工来说，或许还有些神秘。就原始树脂，即漆而言，它的使用只不过就是这样一个过程：将漆倾倒于器物表面，使其均匀的流动，注意避免气泡的出现，仔细的覆盖于器表，最后用灰条、软皮或手推光。那些高档的日本漆器，特别是有一定年份的作品，比如在许多售卖东方器物的店面可以见到的日本清酒杯（图4），明显是经过长时间仔细制作的。漆会上许多层，但依然具有令人惊叹的薄度，甚至，即使破碎，仍无法从漆器断面看到明显的厚度。日本人是极其严谨的，做漆器是这样，其他方面也是如此。然而中国漆器，特别是那些明代制品，尽管少一些工细，却极富艺术精神。在许多实例中，可以发现工匠有意地在第一遍漆干之前髹以第二遍漆，因此出现了那样令人如痴如醉的漆面。这类器物的漆层数量较少，但漆的厚薄保持一致。这样坚固的表面是日本漆器所难以企及的。

这类漆器作品的制作并没有什么惊天秘密，只需切记在制作过程中保持令人不适的高温与干燥的环境，而非欧洲西部那潮湿寒冷的气候。正因如此，漆器的制作，在排除其他条件的情况下，与中国和日本相比，

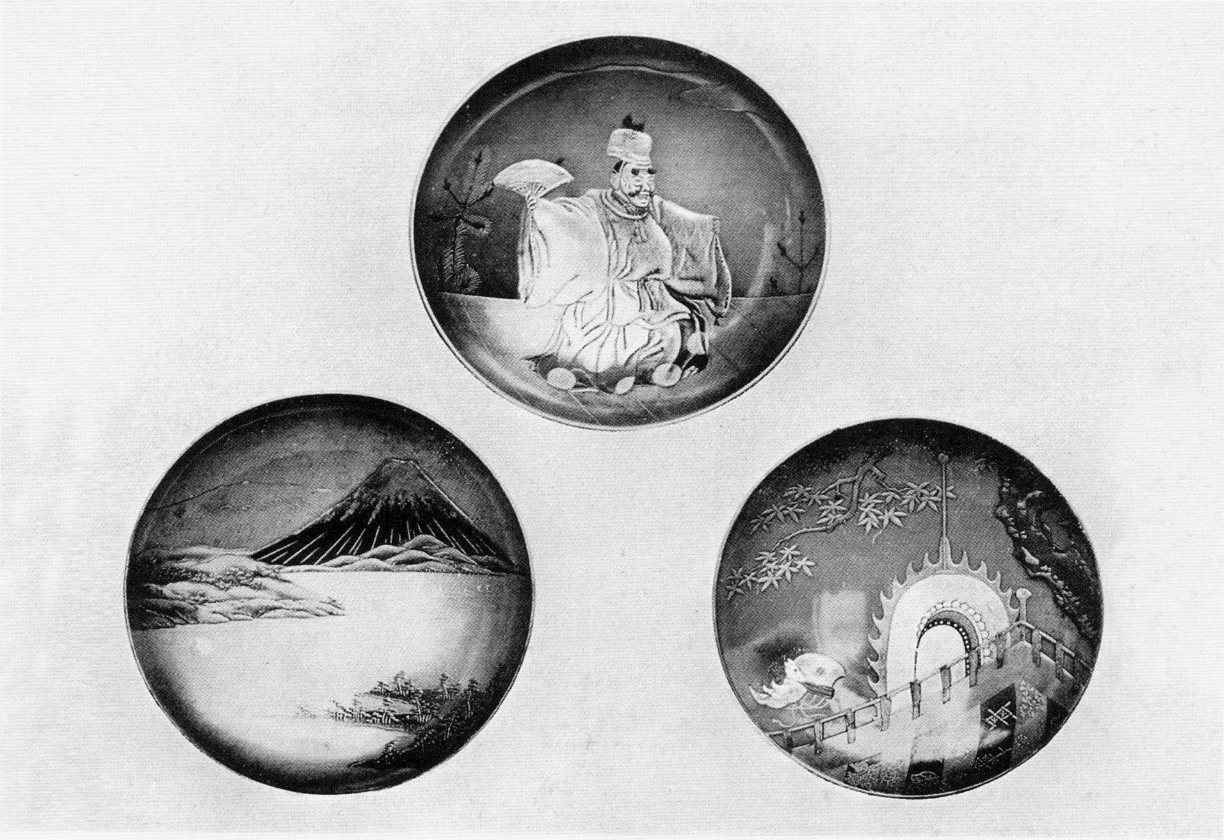

图4

欧洲有着天然的缺陷。在烘干或加热技术出现在西方后，工匠们以这种方法模仿东方的环境来处理漆地。这在中国来说，只能是在条件不允许下的权宜之计。另一点难以为我们所接受的，是东方工匠在制作这些作品中所付出的漫长时间与极大耐心。他们的人工成本似乎微不足道，这只有15世纪前的哥特式木工可与之相较，那个时期食物经常被计入雇佣成本，却是如此便宜（几乎可以忽略）。这种可以忽略不计的成本，恰好是能完成这种大型作品最为重要的因素，比如温彻斯特和切斯特的教堂工程。只有当工资在生产成本中占有一定比例之后，我们才进入到商业时代，此时的艺术创造囿于成本而江河日下。

然而，纵观整个英国手工业发展史，从来没有达到东方艺术那样尽善尽美的水准。即便是18世纪晚期的日本工艺品（此时日本已完成了商业化进程），诸如清酒杯、橱柜、剑锷、坠饰和类似的小物件，依然无可挑剔。一个黑漆或者彩漆的杯子看不出任何加工痕迹，就好像是这些漆是凭空出现在器表的。

许多中国艺术品都要花费数年来完成，寻求那种克服一切困难，将不可能变为可能的喜悦。这种情况并不仅仅存在于真正的艺术品领域，在表面文章或商业价值上也是如此。举一个更为具体的例子来说，在东方市集上经常可以见到的那种镂空雕刻的象牙球，通常有9层或者12层，全部手工雕刻，一层套一层，这种技术与耐心令人称奇；又如玉、水晶或坚石所制香炉，需要采用在表面刻划极细纹饰的玉石雕刻方法；再比如水晶小泪瓶，内径几乎只有⅛英寸，在内部还绘有精致的人物或风景。在西方人看来，只有中国人将自己变作显微镜才可以完成这项工作。对我们来说，很难区分中国艺术品的水准高下，因为它们看起来几乎同样完美。

令西方观念难以理解的还有中国艺术风格的稳定性。在我们的文化中，时尚千变万化，我们可以根据其艺术风格将其定位在一个很短的时期内，误差不过十年，但中国艺术仿佛恒久不变，这超出了我们的概念。当世界上其余地方还是一片荒芜之时，中国已经形成了高度发达的文明。不同的帝国与新的文化起起落落，中国却始终如一。诚然，由于港口开放和中西交流这些因素的影响，中国自明代开始有了一些变化，但依然非常缓慢。在16世纪之前，华人罕见，移民和在中国的外国人也是极个别的，人们对于这个人口众多的庞大国家几乎一无所知。借助我们目前关于中国艺术的粗浅认识，我们可以追溯并标示出它们从汉代到宋代，乃至明代或清康熙、乾隆的变化，但当我们将视野放宽至一千年甚至更长的时间，由于艺术品实例的稀缺，变化就会非常显著。对比中西方古代艺术，就好比缓慢崛起的庞然大物与不断发出嗡嗡声的昆虫。

当然，在一定的条件下，中国人也能够调整自己，来迅速地适应环境。当东印度人装载着茶叶和辣椒，并将本地艺术品诸如咖喱、甚至仅是压舱用的物品引入中国时，几乎没有遇到任何来自传统的阻力。当西方时尚需要一些新点子，无论是漆器、木器、金属、玉石，抑或瓷器，中国工匠都能够迅速做出反应。整个18世纪，中国样式深深地影响到了西欧的木工及制陶者，与此同时，中国人也受到了同等程度的影响。那些新奇的物件运往中国，被复制或者加工，我们从东印度公司的装货清单中可以得到很多线索，例如很多漆家具的结构是西方的，但装饰是中国的。这种做法毋庸置疑，欧洲样式经常被全盘模仿，特别是一些法国藏品实例中，家具都是地地道道中国所制，但设计灵感却来自英国、荷兰或者法国。这只可能是欧洲将样式提供给中国制作的结果。

即使是中国人为了迎合西方品位在样式上有所调整，但装饰上仍旧维持中国风格。传统艺术总会发展至高度程式化，这通常是无法避免的，东方风格的那些装饰主题正好佐证了这种倾向。这种倾向导致

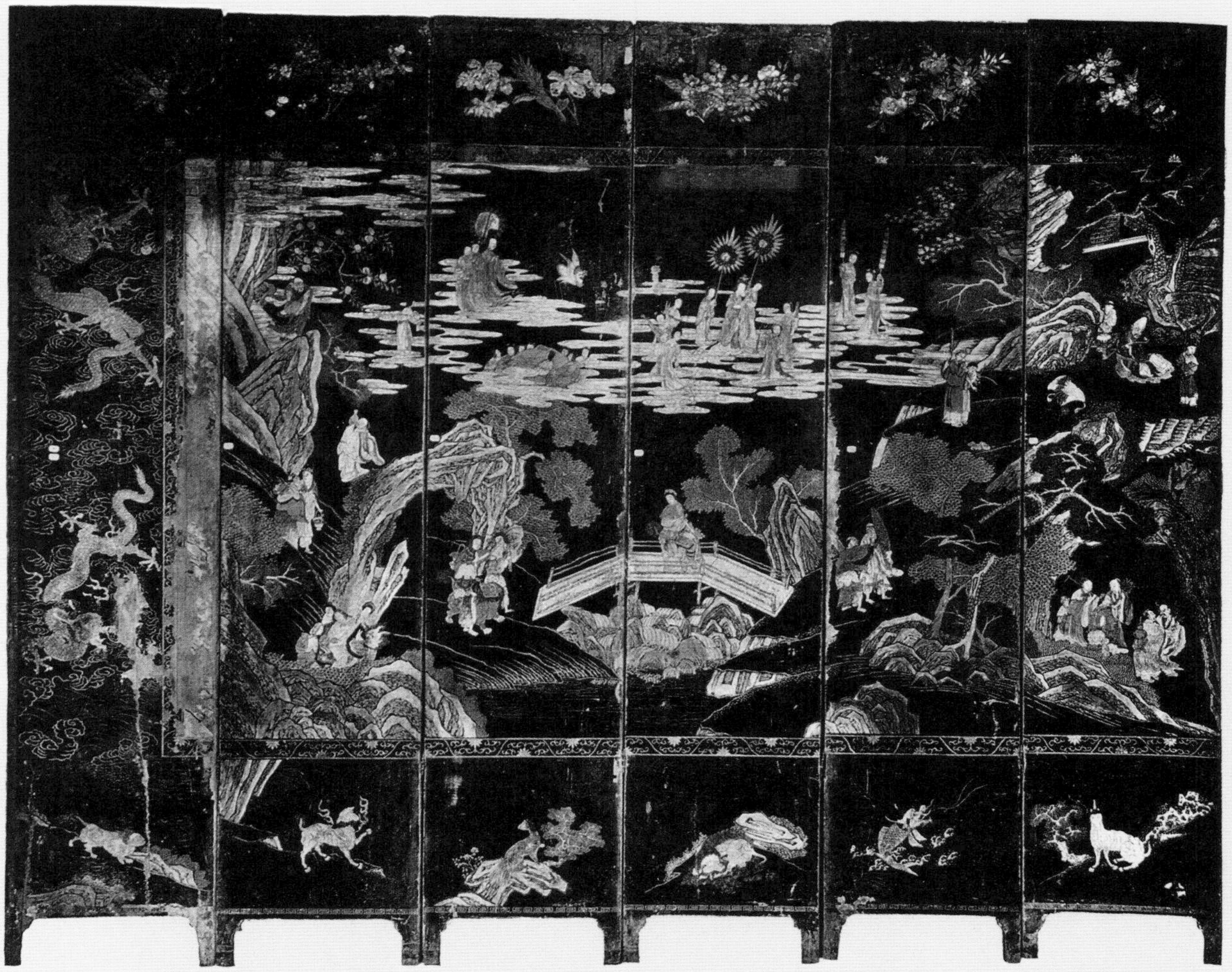

图 5

　　了难以将瓷器和漆器做准确的断代，除非像已故的拉金先生那样，对中国艺术拥有广博的认识及无可匹敌的经验。即便经过漫长的学习，情况依然比较复杂，因为中国的艺术家们会频繁地按照已有的样式与图案仿制。比如，康熙时期就对明代作品仿制得非常频繁，且数量巨大。这不仅要求中国工匠熟知历史，而且要能准确模仿不同朝代的款识，这样的例子在许多漆器上都可以见到。即使那些大型屏风背面有款识，除非可以证实是为中国高官私人订制，否则难以确认其年代或真假。根据已故的威廉·奈尔先生的观点，出于外销目的制作的漆器反而更为安全。

　　另一个能够理解中国设计的重要因素，是几乎所有古代的甚至有些陈旧的艺术品，都将注意力投向实际的装饰而不是造型。中国的艺术家们非常喜欢大的平面或者方形的物件，对于光影并没有煞费苦心的尝试。这是中国艺术与印度斯坦艺术相比，非常显著的一个特点，尽管印度与中国在气候环境上非常相似。塔的样式也许可以作为一个例外，但即使是塔，也更多的是由中国艺术品的模仿者频繁制作，而非中国本土艺术家。中国艺术品尤其是家具，由于规律的缺乏，使得西方人难以理解。这种不理解，加之时间和人工成本的限制，于是便出现了那些浮于表面的拙劣仿品。

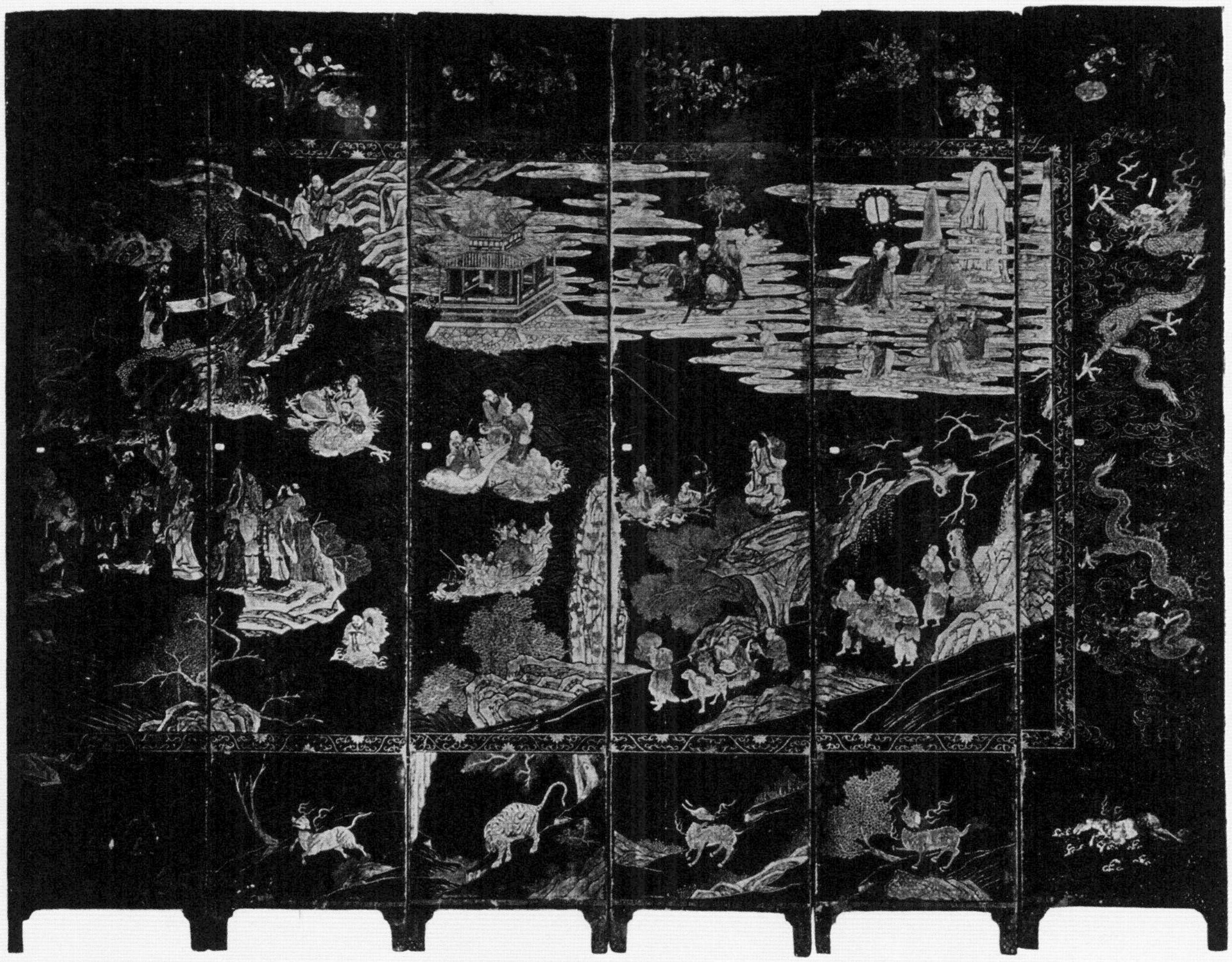

图6

  在欧洲人的观念里，中国的木结构在很多方面是如此的不同。因为我们认为两个金属构件可以通过焊接结合在一起，构件在力上没有损失，而如果是木制榫卯，开槽或者榫接后的力通常会被极大地削弱。这一事实使得我们用材更粗大，或者可以使用那些质地更为坚硬和致密的木材，比如橡木、核桃木或者桃花心木。但是，中国工匠却热衷于使用软木，如同我们喜好中国漆器。对于诸如花瓶或碗托这样的器物来说，并不需要太强的应力，因此工匠们更愿意选择天然木材，尤其是类似花梨木那样颜色和纹理的木材。用软松木来制作家具，其结构通常出奇的复杂，横向或垂直的隔断，最终以精细的榫卯连接，但没有损失力。无论在细节还是整体，都像是毫无疑问的西方构造。在中国潮湿和极端的气候里，木材比在欧洲更容易被彻底处理好，覆盖的漆能够有效的将木胎与空气隔绝。基于此，中国家具，尤其是在漆层没有被破坏的情况下，能够很好地保存至今。

  中国人对于自己所使用漆的保护功能有着很强的信心，这一点超乎寻常。许多高屏风，特别是明代或者清康熙时期的精美实例，如图5到图8，木质也仅仅是软的中国松木而已，两扇竖屏间仅以暗销相连，没有使用任何粘合剂。在欧洲的气候条件下，如果没有厚漆层的保护，这个屏风将在很短的时间内支离

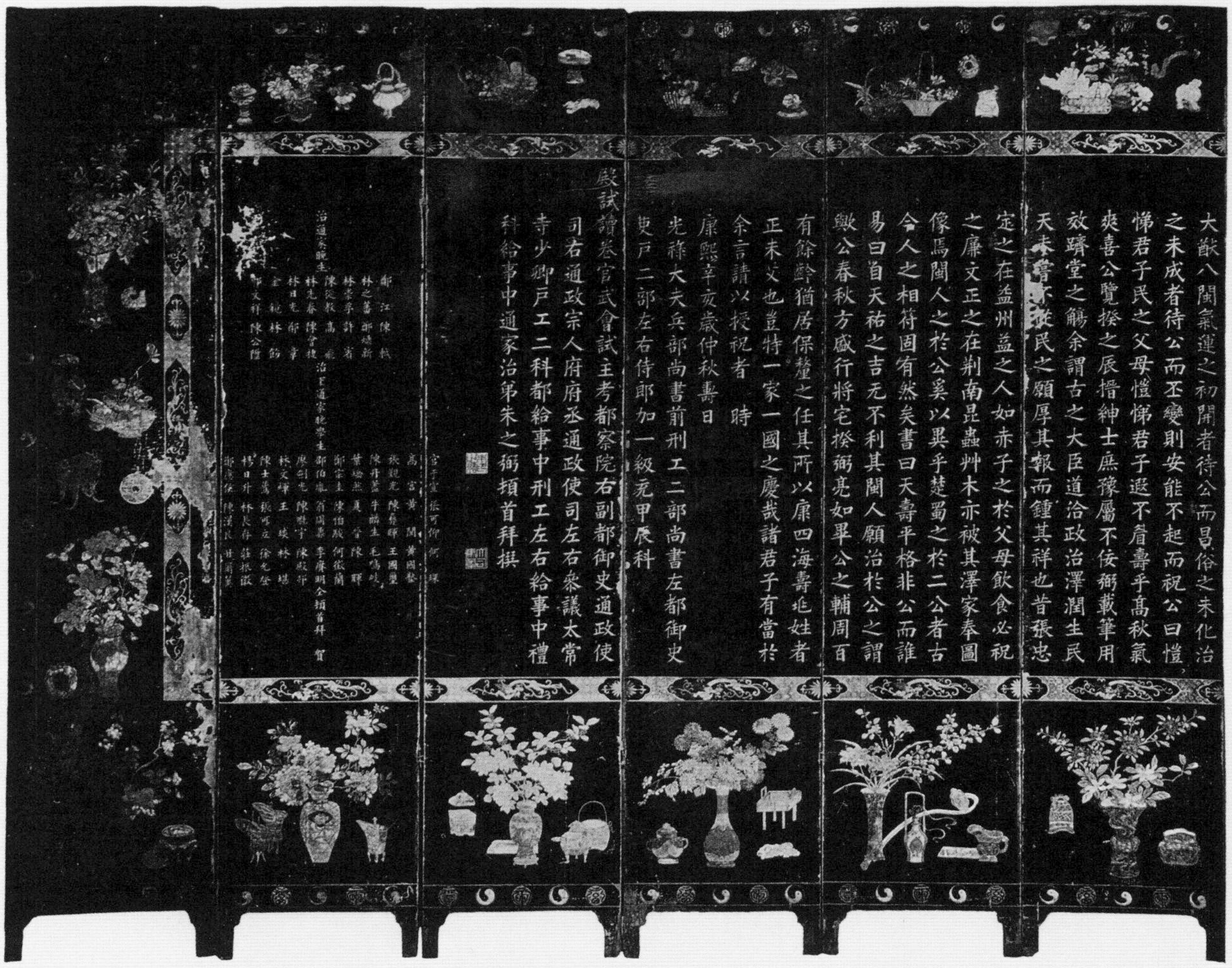

图 7

破碎。而由于这层漆，它从 1671 年，即屏风背面所刻时间，一直保存至今。灰色地上的半透明褐色厚漆层、深雕纹饰以及细灰地，尽管具有明代作品的所有特征，但借助刻在第四扇屏风后面华丽的中文，我们可以得知这件屏风事实上是一位学生送给恩师的礼物，他的名字刻在了最左边，这些证实了屏风是清代制作的。以前以龙爪数量作为判断依据（即五爪龙属于皇帝，四爪龙属于亲王，三爪龙属于一般贵族）的观点，在这一时期，即康熙皇帝统治的前十年里，并不适用。

中国人在呈色的处理上并不总像漆地的处理那样拥有好运气。比如这件屏风，雕漆的图案是在漆地而非木材上，许多英国仿品也是如此。它的做法是雕刻至灰层，然后再上色。屏风上那些重量轻的颜色可以持久保持，但是那些较重的颜色，尤其是朱砂色，因为重量的原因脱落了。金属部件同样也有这种问题。

基于多重因素，特别是由于后代频繁效仿前代，在部分欧洲权威的观点中，对中国家具断代有偏早的趋势。事实上，目前没有太多确切的证据，来将某件作品定到明代，哪怕是 17 世纪初。中国家具在 1500 年至 1720 年间在纯艺术层面达到巅峰，但技术领域的顶点还要晚些，大致自 1725 年直至 18 世纪末。作

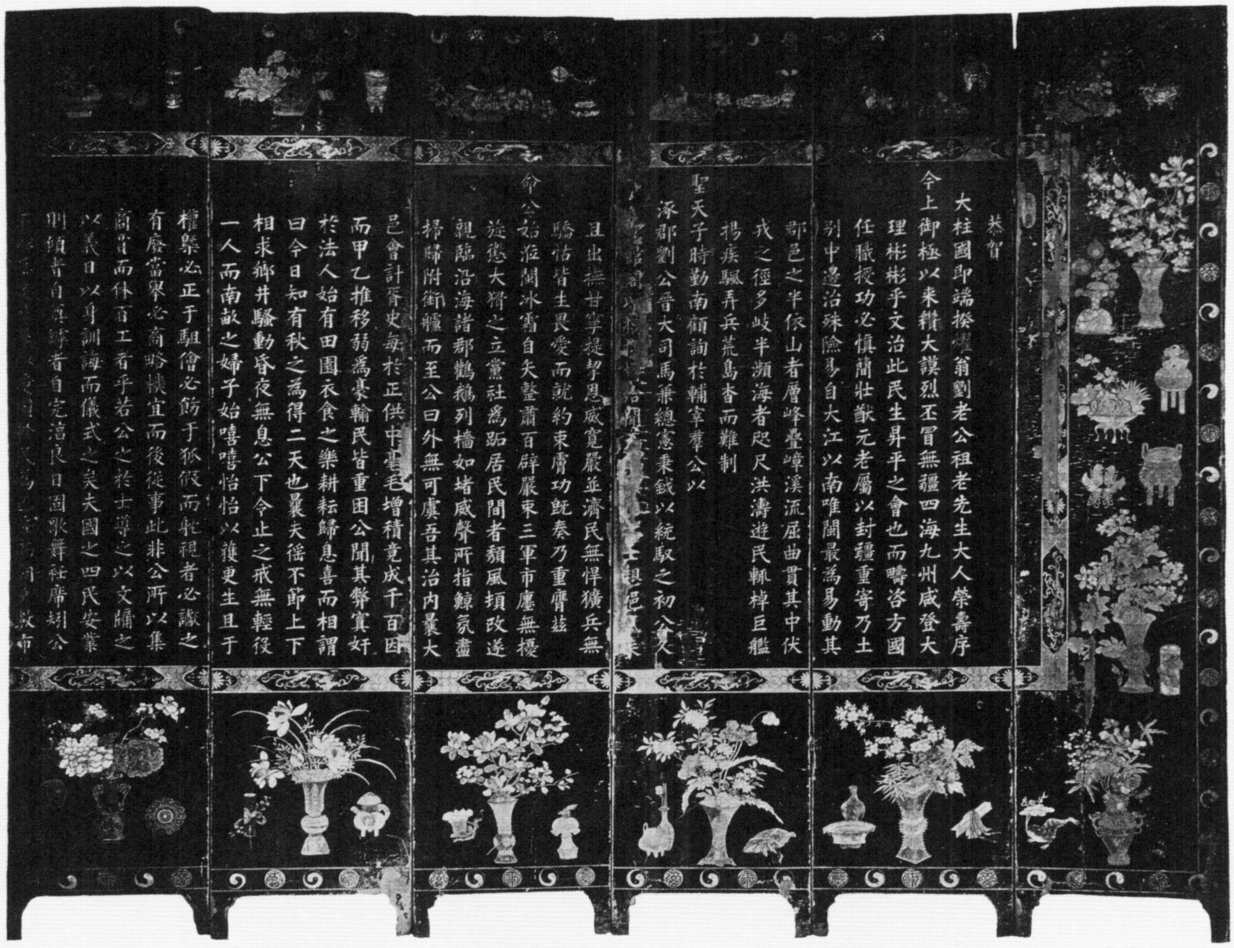

图 8

为普遍准则,这一时期的墨彩和粉彩瓷器也极为出众,尽管绿地粉彩的生产高峰是在 1665 年至 1720 年之间。

比这稍微晚些的漆器作品,最为出彩的当属剔红漆器(有时被误认为"珊瑚"),没有任何一件作品能够超越现存于维多利亚及阿尔伯特博物馆的乾隆时期(1736 ~ 1795 年)剔红宝座,可参见图 9 及图 10。这件作品工艺精美且装饰华丽,存世几无一可与之相较者。

中国和日本都使用这种朱砂漆,而中国自 16 世纪早期便开始制作。明代早期的这种漆器,其红色通常要比清代略深,光的亮度也要高。这两个特征很可能是来自时间的痕迹而非最初的设计。即使在早期尚未使用朱砂的作品中,都没有发现明显的黑色。在日本漆器作品实例中,这种重朱砂以最合适的数量作为添加剂混入,从而保证了即使暴露在阳光下,漆器也不会变成棕色或黑色。与之相对,清朝的朱红漆器在同样的阳光下会褪色而非更深。尽管日本剔红漆器在与中国漆器的竞争中一直略逊一筹,但在光泽度及程式化方面都要高于中国明代及清代作品。也存在这样一种工艺,在漆层通常会嵌入其他颜料或者水晶,乃至各类玉石,使得轮廓更为分明。这种工艺上,日本将自己视作无与伦比的模仿者,但就其

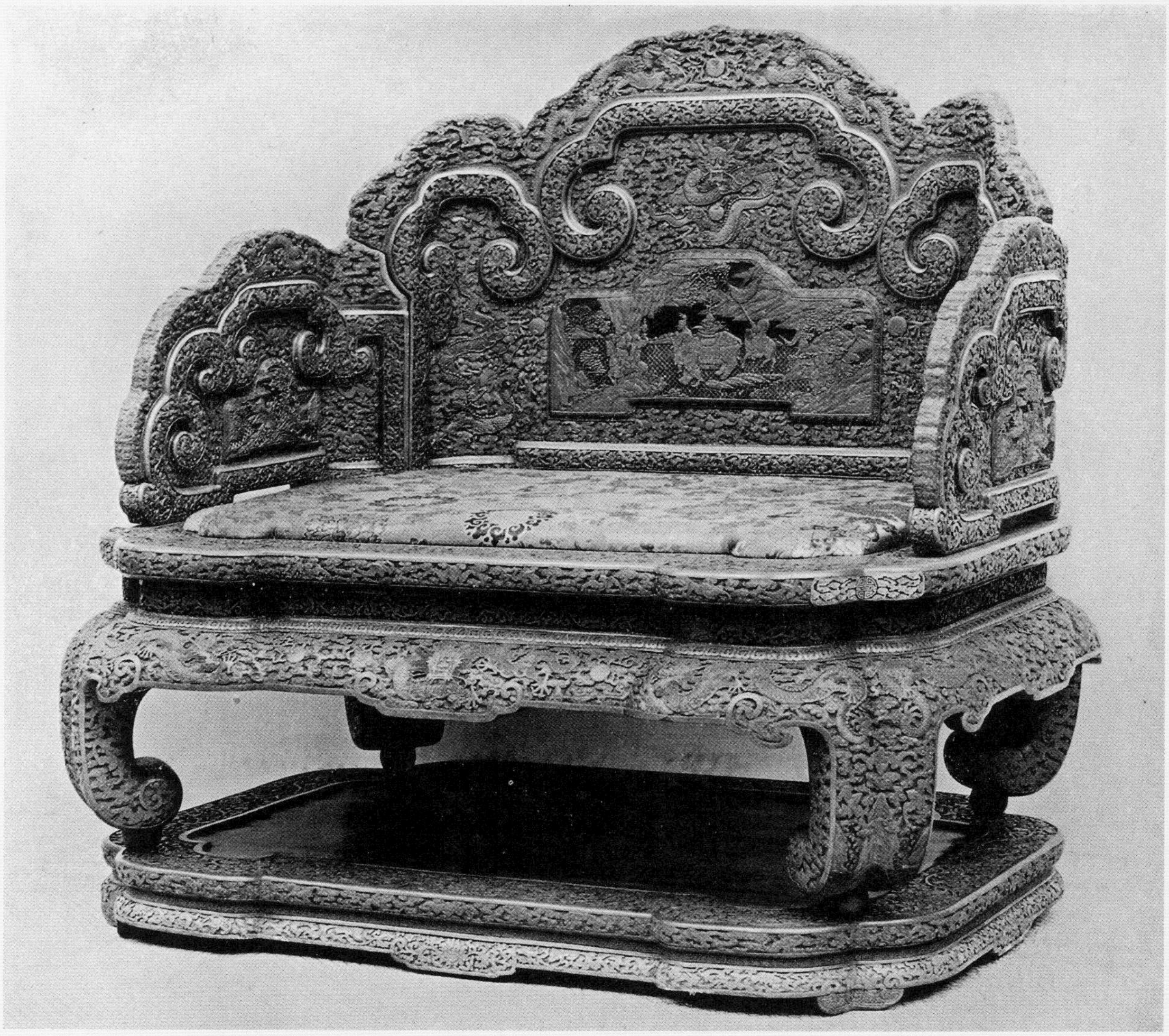

图 9

早期作品来看，也许将其定为创造者更为合适。

图 10 展现了乾隆宝座的一个扶手，中国的榫卯方式前文已有所赘述，通过此图可得进一步了解。

本书中所有图版所示作品皆为彻头彻尾的中国手工制作，但许多实例的样式灵感，可认为来自欧洲，甚至断定为英国。以东印度公司作为媒介的东西方贸易，带来的富有教育意义的影响在这些作品上清晰可见。本书的图版 1 即是具有法国样式的作品，上锁的结构具有明显的 17 世纪中期欧洲风格。由于双开门的问题，包括左边门上锁的做法，在英国或法国只有到了 18 世纪才得以解决，不会更早。中国解决这一问题的方法是使用两个黄铜合页，门的两边各有半个面叶，面叶上有锁鼻用来穿销。在西方观念看来，这样过于明显，甚至有点粗鲁。本书图版 2 作品左门上部的中国人物显示了这个橱柜的时代要早于清朝。

本书图版 5 和图版 6 所展示的橱柜，在很多细节上有相似之处，比如两门之间有垂直的立柱，立柱

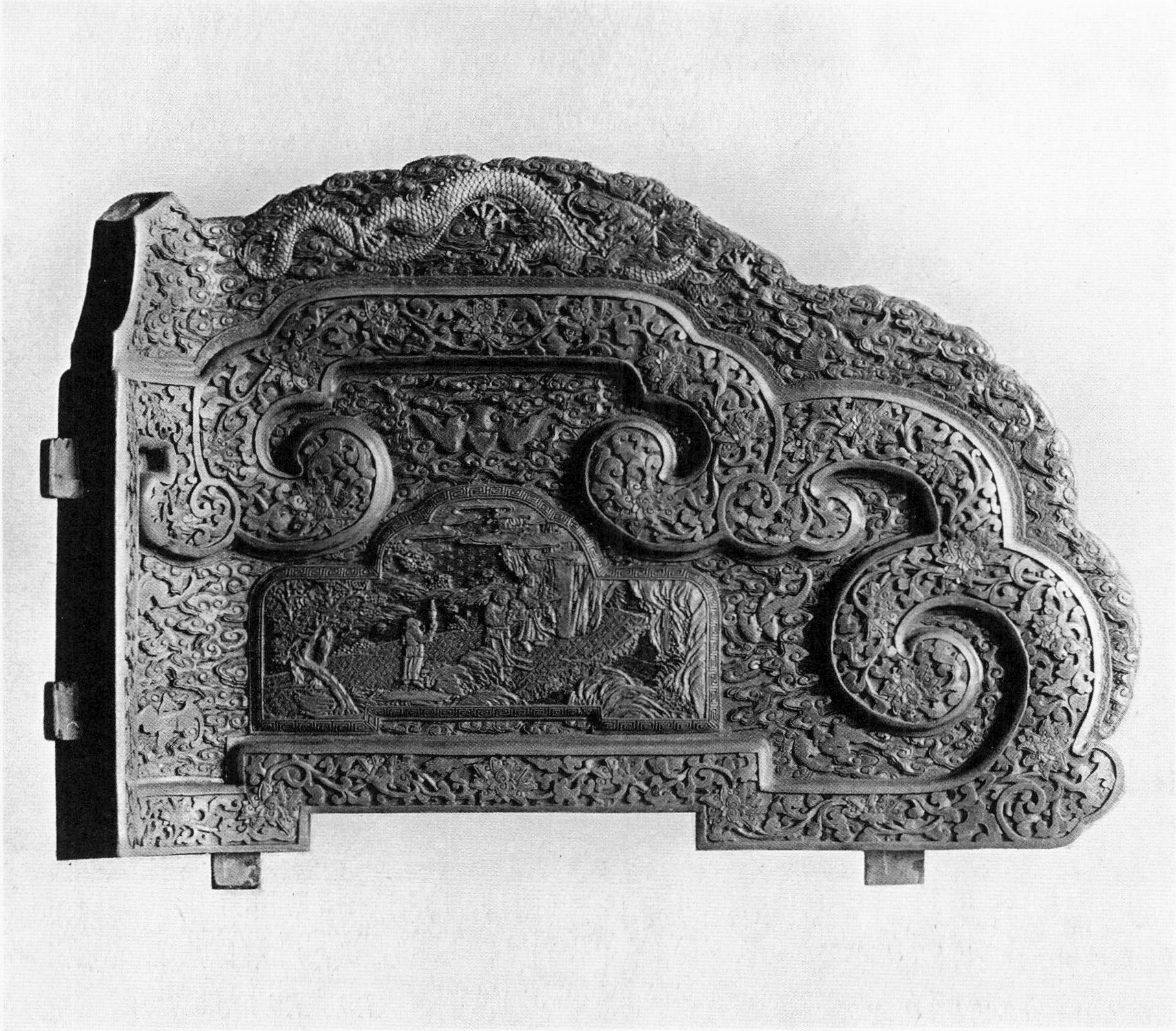

图 10

上安有第三个锁鼻，这样可以保证柜门关上并插上销后能够保持紧闭。后面有很多图版所示家具也有类似的细节。

本书图版 8 所示家具有很明显的清代早期风格，细节粗放且门板厚重。在柜架的周边和挖出的底座牙板处起线，已经有一些对于光影效果的尝试，脱离了明代传统风格，显示出 18 世纪特征。

清代早期所制作的桌椅很明显的受到了西方影响。如图版 23 至图版 26，都是基于英国样式，而图版 26 毫无疑问了受到了荷兰的影响。图版 27 的扶手椅的背板式样是典型的安妮皇后时期样式，类似的实例还可参见图版 41。图版 45 右图的椅子，融合了 17 世纪英国椅子横枨连接的样式，同时也关注到了台座较低的问题，靠背板则汲取了齐本德尔流派的风格。当然其核心风格既不是齐本德尔，也非马提亚·达利，还是来源于 18 世纪中期设计类图书中的中式风格。同样的，图版 42 左图这件作品，也是明显借鉴了荷兰样式的中国家具，这点每一个家具学徒都可以看出来。

本书图版46至图版48所展示的鼓凳或者几，是典型的中国风格。这种样式在几个世纪以来一直维持不变，哪怕是作为一件家具，还是装饰物，几乎不做任何改变。这不仅是漆器，在陶瓷领域也是如此。

聪明的学生应当懂得中国家具的造型装饰有很多值得学习的地方。对于繁忙的商业设计师而言，也许不会即刻被触动，但是一门历经岁月沉淀的艺术不会在匆忙一瞥中便被赞赏。同样的，我们也应该忽略掉一个绘图员那些关于英国风格的零散知识，特别是他将那些混搭成一种不知所谓的时尚。中国风格的设计师将自己投身于艺术，来学习不同时期的作品，区分清代和更早时期的艺术风格，而掌握这些需要经历漫漫的探索之路。对于想学习东方样式的人们而言，中国不同时期的艺术已经很容易混淆，更不要提能够区分中国、日本及韩国各自的艺术作品。理解这种缓慢发展的东方艺术需要经过繁杂的学习过程，但那些付出时间的人终将得到回报。即使开始难以理解，然必将会在某种程度上开阔和完善其艺术理念。而那些将目光仅仅局限于西方艺术自身的人，某种意义上恐是一无所成。

赫伯特·塞斯辛基

1922年8月

INTRODUCTION

It is only within recent years that any exact knowledge has been acquired on the subject of the domestic arts of China, and this is still woefully incomplete. We have had the porcelains and lacquered pieces, the embroideries, carpets and rugs and other products from Pekin, Nankin, Canton. Foochow and Amoy, in Europe, since the days of John Company and the Dutch East India Traders, or even before the ocean route to the East, via the Cape, had been discovered. Wares from the Orient had found their way to the West via the republican trading cities of Venice and Genoa long prior even to the reign of Elizabeth. The charter of the English East India Trading Company was granted in 1600, and trade with the East became regular and extensive, but so little was known of the arts of China or Japan that the products of the two countries (always essentially different to the Oriental) were frequently confused. Thus "The East India Trade, a true narration on divers Ports in East India," issued in 1641, refers to Japan as exporting "all kinds of lackwork," and also states that "the commodities which usually are brought home into Europe are chests, cots, cabinets, and all manner of their householdstuffe."

Of China itself, there occurs hardly a mention in these old records, beyond a brief reference here and there, to the Empire of Cathay.* Japan figures extensively, in fact it gives the name of "japanning" or " jappan work" to pieces of lacquer, and other examples are named, not after their places of origin, but are known by the names of the trading stations from whence they were dispatched to Europe. Thus " Bantam-work," the usual early name for cut or incised lacquer (nearly always of Chinese or Korean origin) commemorates an old Dutch trading station in Java, which was only abandoned as late as 1817 in favour of Sirang, some nine miles inland.

While little or nothing was known of this oriental lacquer-work beyond an acquaintance with the actual pieces in this country, such as large screens and the familiar square cabinets which English workmen mounted on carved and gilt stands, this did not prevent certain charlatans from posing as professors of the art of " lackering " or " japaning." John Stalker " of the Golden Ball " and George Parker of Oxford, published a folio, in 1688, wherein, after all the secrets of the art are revealed,

* Probably from the Russian KHITÂÏ.

the " Epistle to the Reader and Practitioner" states "by these means the Nobility and Gentry (what utter snobs these early bogus professors were !) might be compleatly furnisht with whole sets of Jappan Work, whereas otherwise they were forc't to content themselves with perhaps a Skreen, a Dressing Box or Drinking Bowl, or some odd thing that had not a fellow to answer it."

This book illustrates some of these " odd things," but before referring to the plates in some detail, a few words regarding the lacquer arts of China and Japan, and especially as to the differences between the work of the two countries, may be of some service here.

The Eastern lac is the gum of a native tree (Tsi) *the Rhus Vernificera*, a variety of the Sumach. When freshly exuded, it is soluble, but when dry it is unaffected by even such a drastic medium as alcohol. The European lacquer, which professed to imitate it, is simply slightly glorified coach painting, at its best, and at its worst a mere daubing of paint, surfaced with shellac polish, applied either with a brush or the rubber. In some rare examples, such as the table plateau, Figs. 1 and 2, a degree of perfection is attained, in what is, unmistakably European lacquer-work, almost rivalling that of China or Japan, but even this, fine as it is, cannot be described as true lacquer in the Oriental sense.

There is one cardinal distinction between the furniture of China and Japan, which may be pointed out, here, with advantage, and Fig. 3, one of the so-called Chinese cabinets on an English carved and gilt stand may be cited as typical and explanatory. Many of the so-called Chinese cabinets, such as this, are really Japanese. The latter have, nearly always, one marked characteristic which the former do not possess. The base of the cabinet, proper, consists either of a squat bracket-plinth or stump feet. It was made to stand on the floor, not to be raised some three feet above it. As a general rule, apron-pieces have been fixed behind these stump feet to give an appearance of solidity to the cabinet base at the time when the stands were made.

It is in just this point that the first distinction between Chinese and Japanese furniture can be noted. The Chinaman sits on a chair or stool, the Japanese on the floor. For the latter, every article is dwarfed in proportion. His sitting existence is on a plane three feet lower than that of the Chinese mandarin or notable.

There are other important differences in the lacquer work of the two countries which can only be fully appreciated after an intimate acquaintance with the products of each. The earlier this lacquer is, the more difficult it is to understand, and for one obvious reason. In the same way as the Early English oak often acquires a gloss or polish—usually known as a " patine "—the result of centuries of handling and gentle friction, so time has given a subdued tone and evenness of surface to the Chinese lacquer of the same period, which we know cannot be originally intrinsic. There is little mystery to the practised decorator in modern Japanese lacquer. Granted the original gum — "Tsi" — it is merely a matter of flowing or pouring on evenly, avoiding air bubbles, carefully felting down each successive coat, and polishing the last with a fine powder such as Tripoli, and a soft leather or the fingers. With high grade Japanese lacquer, especially when it is of some age, as in the case of the saki cups which can be found in many Oriental sale rooms (see Fig. 4), there are evidences of long and careful preparation. Many coats are applied, yet the covering is so surprisingly thin, in the aggregate, that when fractured, the lacquer shows no appreciable thickness. The Japanese are mechanical, in this, as in many other things, but they are surprisingly thorough. With the Chinese lac, especially that of the long Ming dynasty, there is much less mechanical perfection, but infinitely more artistic spirit. So crazed is the surface, in many instances, that it would appear as if this were intentional, due to the purposed application of a second coating before the first was dry. The coats are also much fewer in number, and the lac is of thicker consistency. There is a solidity in the ground which is entirely absent in the Japanese work.

While there is no great secret in the application of these lacquer grounds, it must be remembered that the work is done under far more favourable conditions in a hot and dry climate, than in the humid and chilly air of Western Europe. For this reason alone, even if none other existed, the European is seriously handicapped when placed in artistic competition with the Chinese or Japanese. In the West, it is necessary to imitate the Oriental conditions by stoving or heating the lacquer grounds after they are applied, and this is only a poor substitute for the conditions prevailing, say, in China. Another point which cannot be readily appreciated in this country is the infinite time and patience taken by the Oriental in his work. His living cost is so trifling, that it is only comparable with that of the Gothic woodworker prior to the fifteenth century, a period when food was so cheap that it was often included (by way of an inconsiderable afterthought) in the terms of hiring. This negligible cost of living was one of the most powerful factors in the production of such magnificent works as the choir stall canopies at Winchester and Chester, or the decorated chancel screens at Bramfield, Ludham or Ranworth. It is only when wages became to be an appreciable item in the cost of production that we arrive at the age of commercialism. When artistic creations have to be lopped and stultified on the score of expense.

Never, however, in the entire history of English handicrafts, has the patient attention to detail or finish of the Oriental been even approached. Even in the Japanese work of the late eighteenth century (Japan has been commercialised since) in saki cups, tiny cabinets, tsuba, netsuké and similar small objects, the finish everywhere is superb. A cup in black or coloured lacqner bears no sign of contact anywhere while the work was in progress. It was as if the piece were lacquered while invisbily suspended in the air.

Many of the Chinese productions must have taken many years to accomplish, and are always monuments of incredible difficulties attempted and overcome as if for the sheer pleasure of doing the impossible. This is the case not only with real art works made for mandarins, but with articles of nominal or commercial value only. As specific examples may be mentioned the pierced and carved ivory balls which are general in Eastern bazaars, where as many as nine or twelve are nested the one inside the other, all carved, the one through the piercings of the other, with marvellous skill and patience, from the one piece; the *koro* in jade, crystal or other hard stones, where to even scratch the surface demands the methods of the lapidary, or the tiny crystal tear bottles, hollowed inside evenly through an aperture often less than one eighth of an inch in diameter. and then painted with tiny figures or scenes, *from the inside*. It would almost seem to the Occidental as if the Chinaman could reduce himself to microscopical proportions for such work. To our eyes there is little difference in mere quality between Chinese work of the highest or the lowest grade; the perfection of finish is still equally marvellous in them all.

Just as incomprehensible to our Western ideas is the fixity of the conditions in China. In our civilisations, where fashions change with kaleidoscopic rapidity, in such short periods of time that we can often dare an art work, by its style alone, within a margin as narrow as a single decade, the apparently never-changing character of Chinese art is beyond our conception. China was civilised, and highly developed, in th ages when the rest of the world was barbaric, yet has remained almost stagnant while other empires and new arts have arisen, flourished and fallen to decay. True, China has also grown since the days of the long Ming dynasty, but by slow degrees, although hastened by such factors from without as the opening of trading ports and intercourse with the West. During the sixteenth century, and before, the Chinaman rarely, if ever emigrated, and foreigners were exceptional, if not almost unknown, within the frontiers of this enormous and densely populated country. With our present knowledge of Chinese arts, perfunctory as it is, we can look back and mark the changes from Han to Sung, to Ming, to K'hang-hsi and Ch'ien-lung, but we are looking through a perspective of a thousand years and more, and the changes are notable simply by reason of their paucity in numbers. Compared with the developments of Western Arts it is as the slow upheaval of a leviathan compared to the buzzing of a gnat.

Yet, given certain conditions, the Chinese can change, and can adapt themselves quickly to altered conditions. In the days when the East Indiamen loaded with their tea and spices, using products of their domestic arts merely as curios, or even as ballast, there was little incentive to struggle out of the age-old grooves. When Western fashions demanded something new from the Chinese worker in lacquer, wood, metal, hard stones or porcelain, it was rapidly forthcoming. If, during the eighteenth century, the Chinese forms influenced the woodworker and the potter of Western Europe, it is equally certain that the Chinaman was affected, in almost equal degree. That pieces were sent to china, to be copied or to be finished, we know from many of the East India Company's bills of lading, and we have many examples of lacquered furniture where the construction is European and the decoration Chinese. What is not recorded in such unmistakable manner is the fact that the European models were often copied throughout, and articles of furniture are to be found, principally in French collections, where everything is Chinese excepting the designing inspiration, and that is unmistakably English, Dutch or French. This can only have been the result of a demand from Europe being supplied from China.

If the Chinese altered their models as a concession to Western tastes, the ornament remained unaffected. It is inevitable that an age-old art should become highly conventionalised and Oriental decorative motives exemplify the truth of this in a very marked degree. It is this which renders much of the porcelain and nearly all the lacquer work so difficult to date, unless one has the unrivalled experience of the late F.J.Larkin, for example, one whose knowledge of Chinese art was profound. To make a difficult study still more complicated, the Chinese artists frequently forged both the designs and the marks of the models from the older dynasties. Thus, forgeries of Ming pieces during the reign of K'hang-hsi are not only not infrequent; they abound. It requires a cultured Chinaman not only to read, but to be able to vouch for the accuracy of many of the signatures and histories which are to be found on many pieces of lacquer, especially on the reverse of the large screens, unless it can be proven that such pieces were made for the private use of some high mandarin. With lacquer work intended for export, it is safer to adopt the attitude of the late William Nye.

Another important factor necessary for the comprehension of Chinese designing, one which is true of nearly all very ancient and somewhat stagnant arts, is that form receives far less attention than actual decoration. The Chinese artists content themselves principally with large flat surfaces or square objects; it is rare to find any elaborate attempts at light and shade. It is in this characteristic that the Chinese differ so strikingly from the Hindustani, although in both India and China the climatic conditions are similar. Perhaps one exception should be made in favour of the pagoda form, but even here the pagoda is much more frequently used by imitators of Chinese work than by Chinese artists themselves. It is this comparative absence of play of line which renders Chinese art, especially in furniture, so difficult to grasp by the Westerner; there is nothing to take hold of, and time and cost of labour forbid any other than a feeble imitation of the actual surface decoration.

Chinese construction, in wood, differs also, in many important respects, from the European. In metal, two pieces can be soldered or welded together at the joint, with no loss in strength to either piece, but the tenon or groove in wood must weaken the piece which is grooved or mortised, often very materially. This fact necessitates the use of timber of larger scantling than would, otherwise, be required, and timber of hard or tough texture is also selected for the purpose, whether oak, walnut or mahogany. The Chinese use a soft wood, very similar to our popular for their lacquer work, although for some pieces, even where great tensile strength is not demanded, such as vase or bowl stands, they nearly always select a native wood, in colour and texture akin to rosewood. In furniture made from the soft white pine, their construction is often exceedingly complex. A lateral shelf or vertical partition is housed into an end with a number of very small mortises and tenons which do not weaken, either in detail or aggregate, as Western construction undoubtedly does. The wood itself is more thoroughly seasoned than is possible in Europe, with its humid and extreme climate, and the covering lacquer effectually seals it from atmospheric effects. It is for this reason alone that Chinese furniture, especially if the surface of the lac be unbroken, stands so well even in this country.

The reliance placed, by the Chinese, on the protective qualities of their lacquer, is extraordinary. Many of the tall screens, especially of the Ming or K'hang-hsi periods, such as the magnificent example illustrated here in Figs. 5 to 8, where the wood is merely a soft Chinese pine, have the panels in two vertical sections, jointed together with dowels, but without any adhesive. Unprotected by the thick covering lacquer, this screen would fall to pieces, in our European climate, in a very short space of time. With its lacquer, it has persisted from 1671, the date inscribed on the reverse side. Although possessing all the characteristics of Ming work, in the thick semi-transparent brown lacquer glazed over a grey ground, with deep incising, grounded with a preparation of fine sand, the inscription on the reverse, setting forth, in flowery Chinese, the fact that the screen was presented to a Professor by his pupils, whose names are figured on the extreme left hand side, proves that the work is Manchu. The significance of the dragon's toes or claws, which had existed before (five for Imperial pieces, four for royal princes and three for mandarins) does not apply at this date, the end of the first decade of the reign of K'hang-hsi.

The Chinese are not always as fortunate with their pigments as with their lacquer grounds. The cut, incised lacquer, such as on this screen, is carved in the preparation, not in the wood, as many of the English imitations are. The ornament thus cut out, is finished with a sanded preparation, on which the final colours are applied. Those pigments which are light in weight have persisted in this screen, whereas the heavier colours, especially the vermilion, have either perished or fallen off by reason of their mere weight. The same fate has overtaken many of the metals which were used.

Owing to a multiplicity of causes, principally to the fact that the work of one dynasty or reign frequently copied that of a previous time, there is always a strong tendency, on the part of European authorities, to date Chinese furniture much too early. Actually, there is very little extant at the present day which can be referred, with any certainty, to the Ming period, even as late as its as its close in the early years of the seventeenth century. The furniture of China reaches its zenith, in pure artistry, in the years from 1500 to 1720, but its highest level of mere mechanical perfection is later than this, from about 1725 to the end of the eighteenth century. The fine porcelains, Famille Noir and Famille Rose date from this period, almost as a general rule, although the Famille Verte of the highest quality was made in the years from 1665 to 1720.

Of this later lacquer work, the highest limit of excellence was reached in the carved cinnabar lac (sometimes erroneously styled " coralline") and no finer example of its kind can be cited than the imperial throne of Ch'ien-lung (1736—1795) now in the Victoria and Albert Museum, and illustrated here in Figs. 9 and 10. For absolute perfection of workmanship and intricacy of ornament, this throne is almost incomparable.

Both the Chinese and the Japanese used this cinnabar lacquer; in the case of the former, from the early years of the sixteenth century. The early Ming carved red lac is usually darker than that of the Manchu dynasty, and is generally more highly polished; two qualities which may be due more to the passage of time to original tone or finish. There is no blackening visible, even in the earliest work, however, which shows that vermilion was not used. In the case of the Japanese lacquer, however, there is usually a strong admixture of this heavy vermilion, and very little, even of the finest quality, will stand exposure to strong sunlight without turning either brown or black, whereas, with the Manchu cinnabar lac the same light will bleach, rather than darken it. The Japanese carved red lac, although inferior to the Chinese in permanence of its composition, is generally much more brilliant and highly modeled than either the Ming or the Manchu work. It is also, as a rule, enriched with other colours or with insets of crystal or other hard stones, finely chiseled. It is in this craft, especially, where the Japanese show themselves as unrivalled imitators, but it is to their earlier work where one must turn to judge them as creators.

In Fig. 10 is shown one of the arms of Ch'ien-lung's throne, and the Chinese constructional methods as outlined in a previous page, can be studied in this illustration.

The examples illustrated in the plates of this book are all of Chinese workmanship throughout, although the inspiration of the general form, in many cases, is decidedly European, and in some positively English. It is in these pieces where the educational influence of the trade of the West with the East, through the media of the East India Companies, is so apparent. Thus, Plate I., is French in type, with doors pin-hinged in the European manner of the mid-seventeenth century. It cannot be earlier, as the problem of double folding doors—which involves the use of bolts on the left hand door—had not been solved in England or France until the eighteenth century had been reached. China has surmounted the difficulty by the double brass plate, one half on each door, fitted with eyelets, skewered through, a palpable, and somewhat crude concession to Western ideas. The Chinese characters on the detail of the upper left-hand door (Plate II.) show that this cupboard is prior to the Manchu dynasty.

In the cupboard shown in Plates V. and VI.— which closely resemble each other in many details—there is a central vertical partition between the doors, to which is fixed a third eyelet, thereby holding the two doors rigid when closed, and with the skewer inserted. Many of the pieces shown in the later plates have a similar detail.

Plate XVIII. is typical Chinese construction of early Manchu type. The details are heavy, and the doors thick. There is some attempt here at aiming at the effect of light and shade, in the rounding of shelf edges and cutting out of the plinth apron, an indication of the eighteenth century, and a departure from the Ming traditions.

It is in the tables and chairs of the early Manchu period where the Western influences are the most apparent. Thus Plates XXIII., XXIV., XXV. and XXVI. are based on English models, and Plate XXVI. is, as unquestionably, Dutch in inspiration. The chair in Plate XXVII. is a typical Queen Anne splat-back pattern, and the same may be said of the two in Plate XLI. The one on the right of Plate XLV. is an amalgamation of a seventeenth century English stretchered chair, as far as its lower stage is concerned, with a back which owes its inspiration to the Chippendale school. It is certainly not the source from which Chippendale, or rather Matthias Darly, copied the Chinese models which figure in the design-books of the middle eighteenth century. Equally, the chair on the left of Plate XLII. is a Chinese adaptation of a Dutch model, of the kind with which every student of furniture is acquainted.

In the vase or cistern stands such as in Plates XLVI., XLVII. and XLVIII., the Chinese are quite individual. These are the models of the country which persist, with very little modification, for centuries, either as pieces of furniture, or as objects of decoration, not only in lacquer but in porcelain also.

There is much to be learned from this Chinese furniture by the diligent student of decorative form. To the busy commercial designer the appeal is not so immediate, but an art which has been refined through ages is not to be appreciated in the rare intervals of a hurried existence. In the same way as we would ignore a draughtsman with a casual knowledge of English styles—especially if he were to mix them together in an ignorant fashion—the designer in Chinese styles owes it to himself and his art to study the work of the various periods, to be able to differentiate between the Manchu and the earlier styles, and to accomplish this a good deal of painstaking research is necessary, especially for those whose acquaintance with Oriental form is so meagre that they confuse not only the Chinese art of the various dynasties but are not able to distinguish between Chinese, Japanese or Korean. It is a complicated study, this slow evolution of Oriental art, but it is one which will repay the student for the time involved, and will help, in a degree which will not be apprehended at the outset, to broaden and to refine his artistic outlook, in a way which the close examination of Western work will utterly fail to do.

HERBERT CESCINSKY.

August, 1922.

图 版
PLATES

图版 1
PLATE I.

柜，黑漆，嵌螺钿
高 205.7 厘米　宽 119.7 厘米　深 57.5 厘米

奥迪隆·罗什先生藏

Chest, black lacquer, mother-of-pearl inlaid and carved. Height, 6 feet 9 inches; breadth, 3 feet 11⅛ inches; depth, 1 foot 10⅝ inches.

In the possession of O. Roche, Esq.

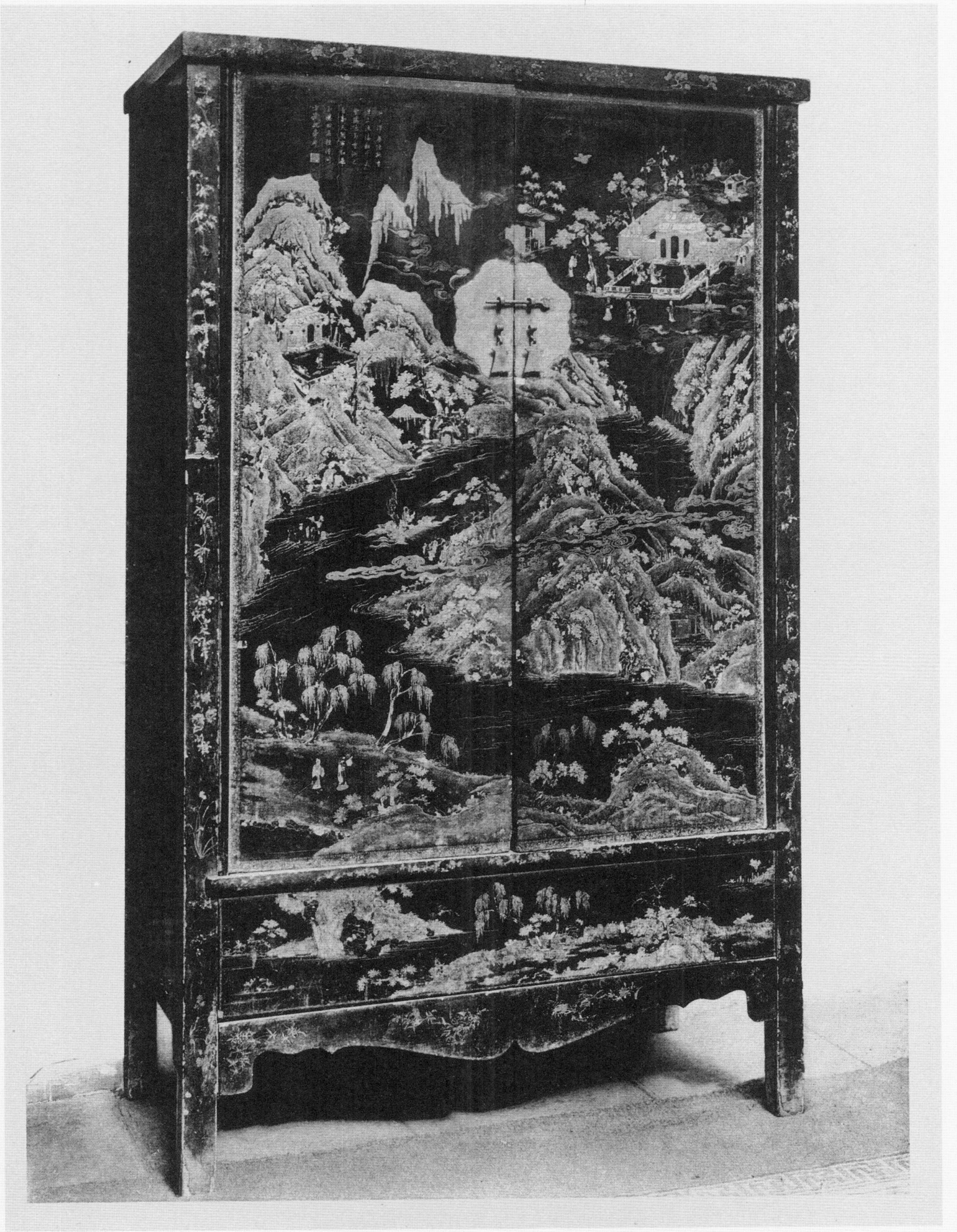

图版 2
PLATE II.

柜（图版 1）的局部
Detail of the chest, Plate I.

图版 2
PLATE II.

胸部之形成， 图版 I.
Decad of the chest, Plate I.

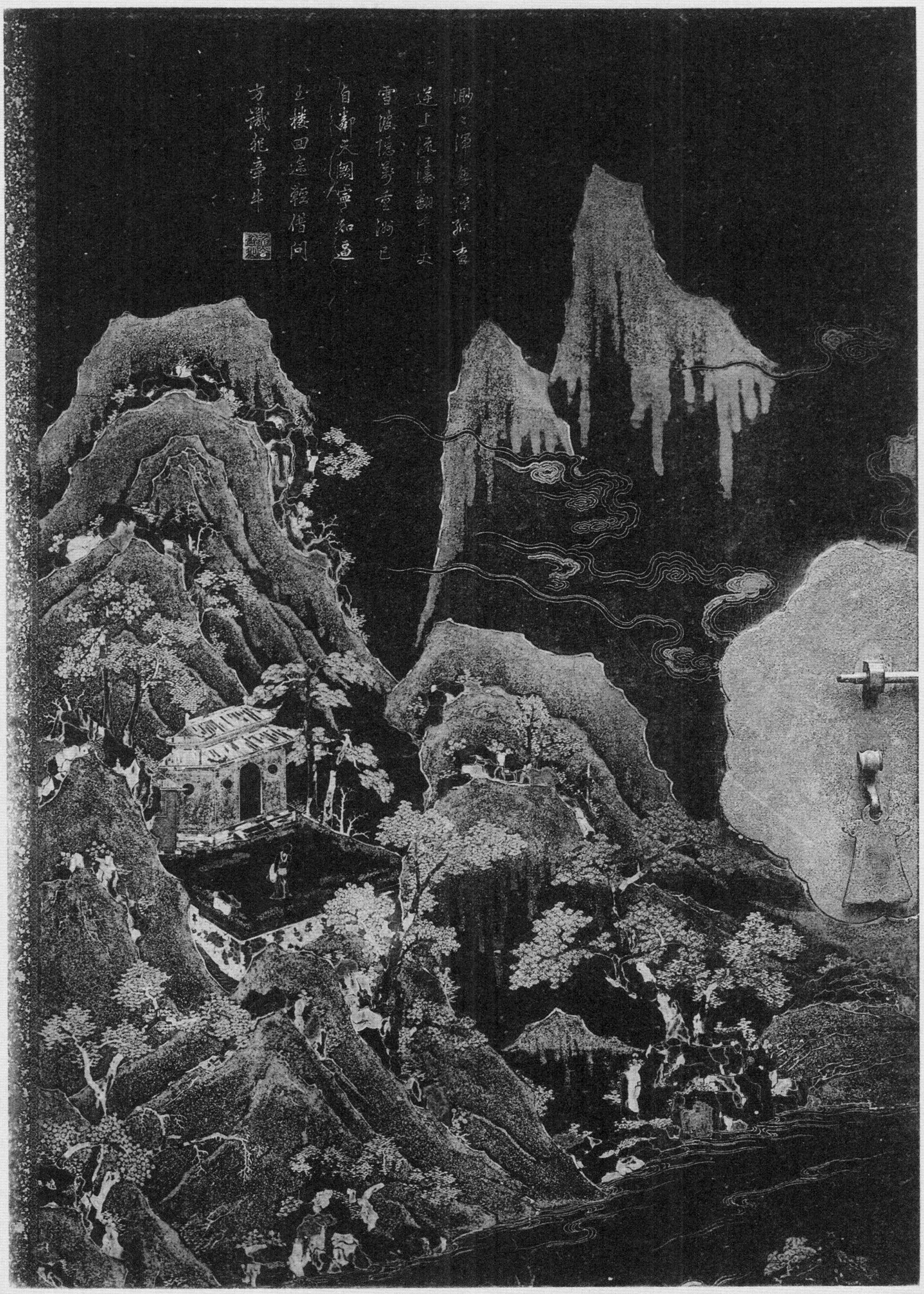

图版 3
PLATE III.

柜（图版1）的局部
Detail of the chest, Plate I.

插圖 3
PLATE III.

梔（圖之1）的局部
Detail of the above, Plate I.

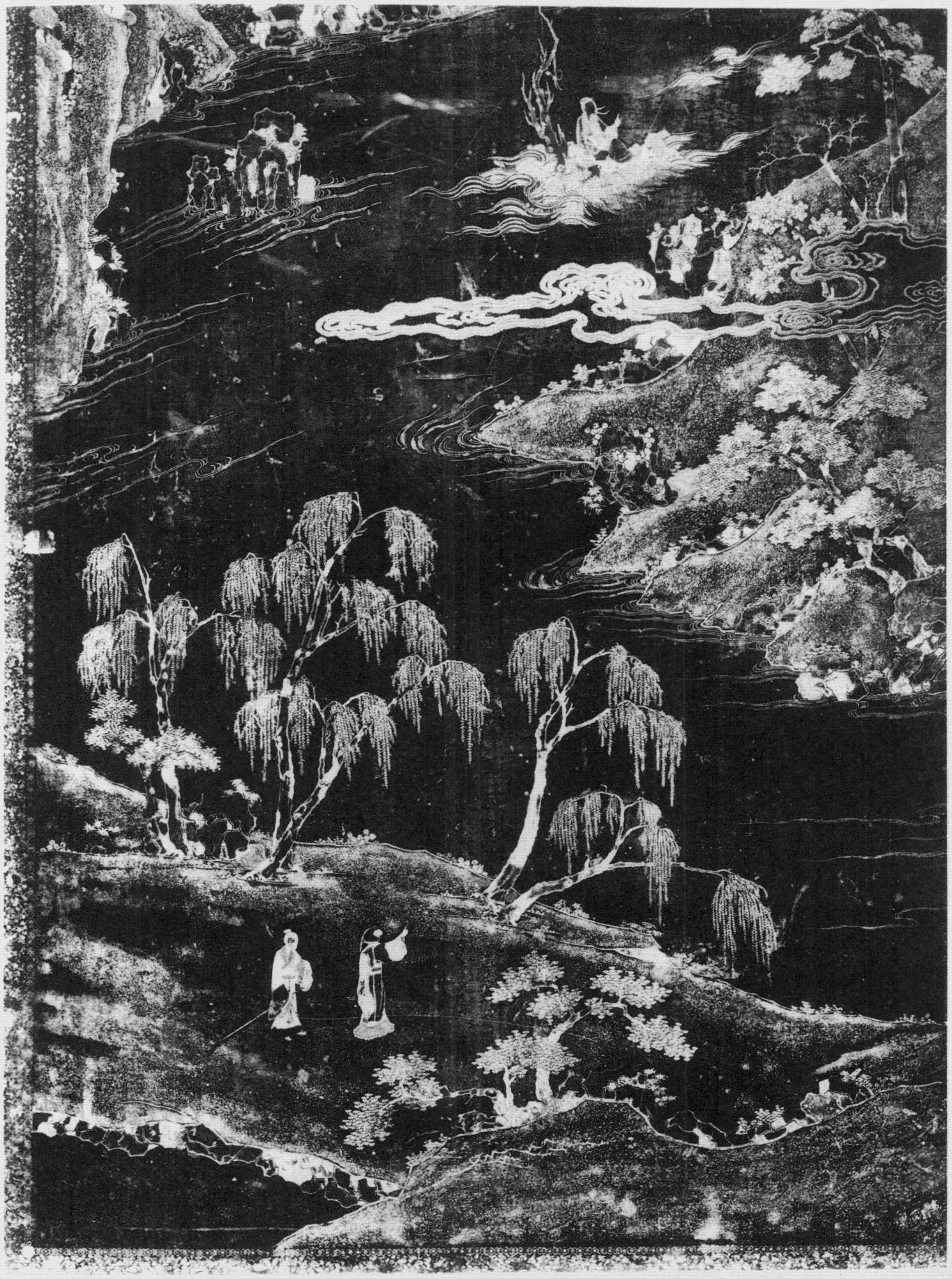

图版 4
PLATE IV.

柜，木质，嵌螺钿
高 193.7 厘米　宽 99.7 厘米　深 61.6 厘米

查尔斯·维涅先生藏

Chest, natural wood, mother-of-pearl inlaid and carved. Height, 6 feet 4¼ inches; breadth, 3 feet 3¼ inches; depth, 2 feet ¼ inch.

In the possession of Charles Vignier, Esq.

图版 5
PLATE V.

顶箱柜，黑漆，浮雕描金
高 169.2 厘米　宽 87 厘米　深 42.5 厘米

奥迪隆·罗什先生藏

Double-chest, black lacquer, gold ornament in relief. Height, 5 feet 6⅝ inches; breadth, 2 feet 10¼ inches; depth, 1 foot 4¾ inches.

In the possession of O. Roche, Esq.

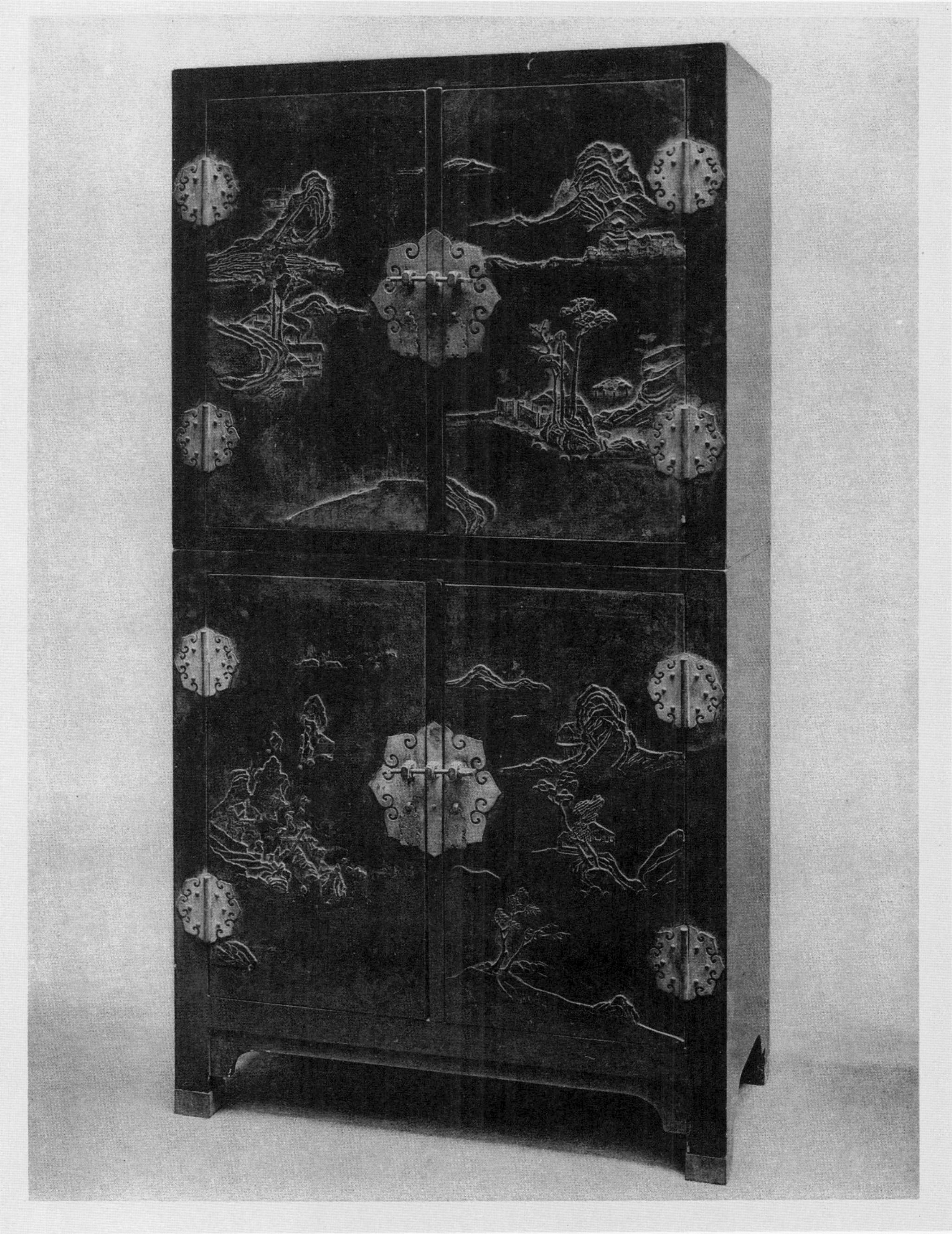

图版 6
PLATE VI.

柜,黑漆,浮雕描金
高 179.1 厘米 宽 182.3 厘米 深 59.4 厘米

奥迪隆·罗什先生藏

Chest, black lacquer, gold ornament in relief. Height, 5 feet 10½ inches; breadth, 5 feet 11¾ inches; depth, 1 foot 11⅜ inches.

In the possession of O. Roche, Esq.

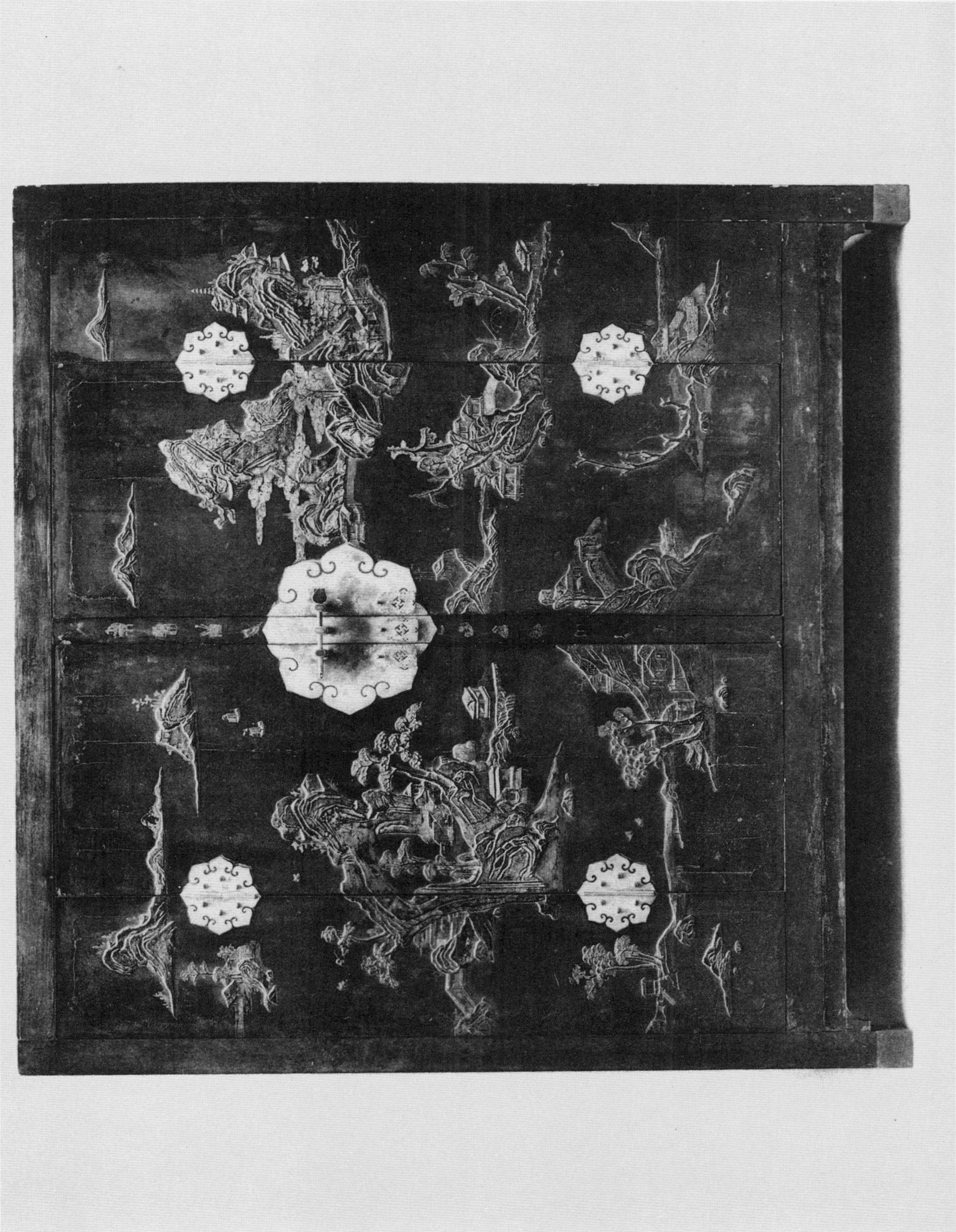

图版 7
PLATE VII.

柜，红漆，彩绘
高 224.8 厘米　宽 175.9 厘米　深 63.5 厘米

沃尔希先生藏

Chest, red lacquer, polychrome decoration. Height, 7 feet $4\frac{1}{2}$ inches; breadth, 5 feet $9\frac{1}{4}$ inches; depth, 2 feet 1 inch.

Worch Collection.

PLATE VII.

Chest, red lacquer, polychrome
decoration. Height, 21/2 of 1 inches;
breadth, 31/2 of 2 inches; depth, 2
feet 1/2 inch.

Brown Collection

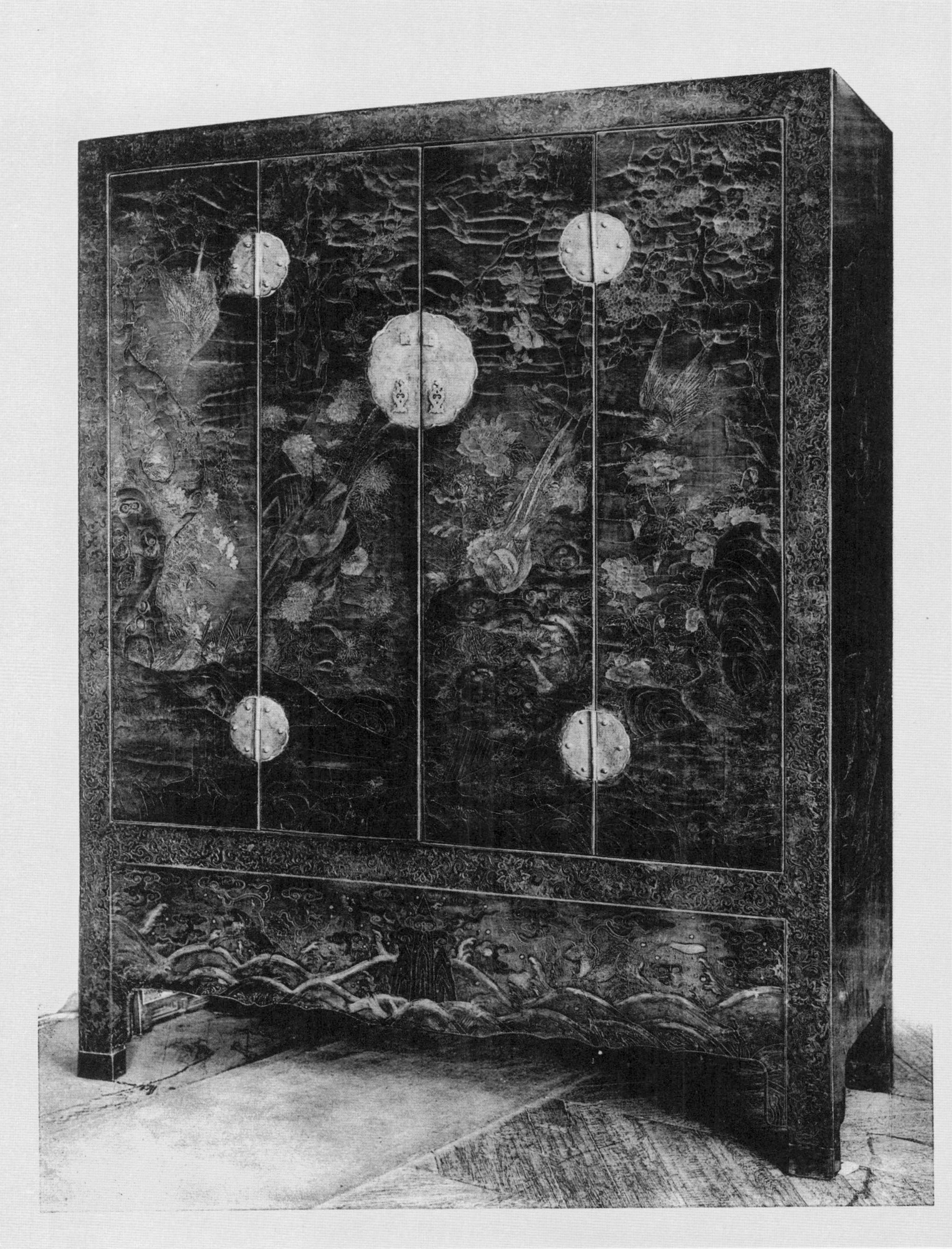

图版 8
PLATE VIII.

顶箱柜，黑漆，嵌螺钿
高 149.2 厘米　宽 88.9 厘米　深 57.5 厘米

奥迪隆·罗什先生藏

Double-chest, black lacquer, mother-of-pearl inlaid and carved. Height, 4 feet 10¾ inches; breadth, 2 feet 11 inches; depth, 1 foot 10⅝ inches.

In the possession of O. Roche, Esq.

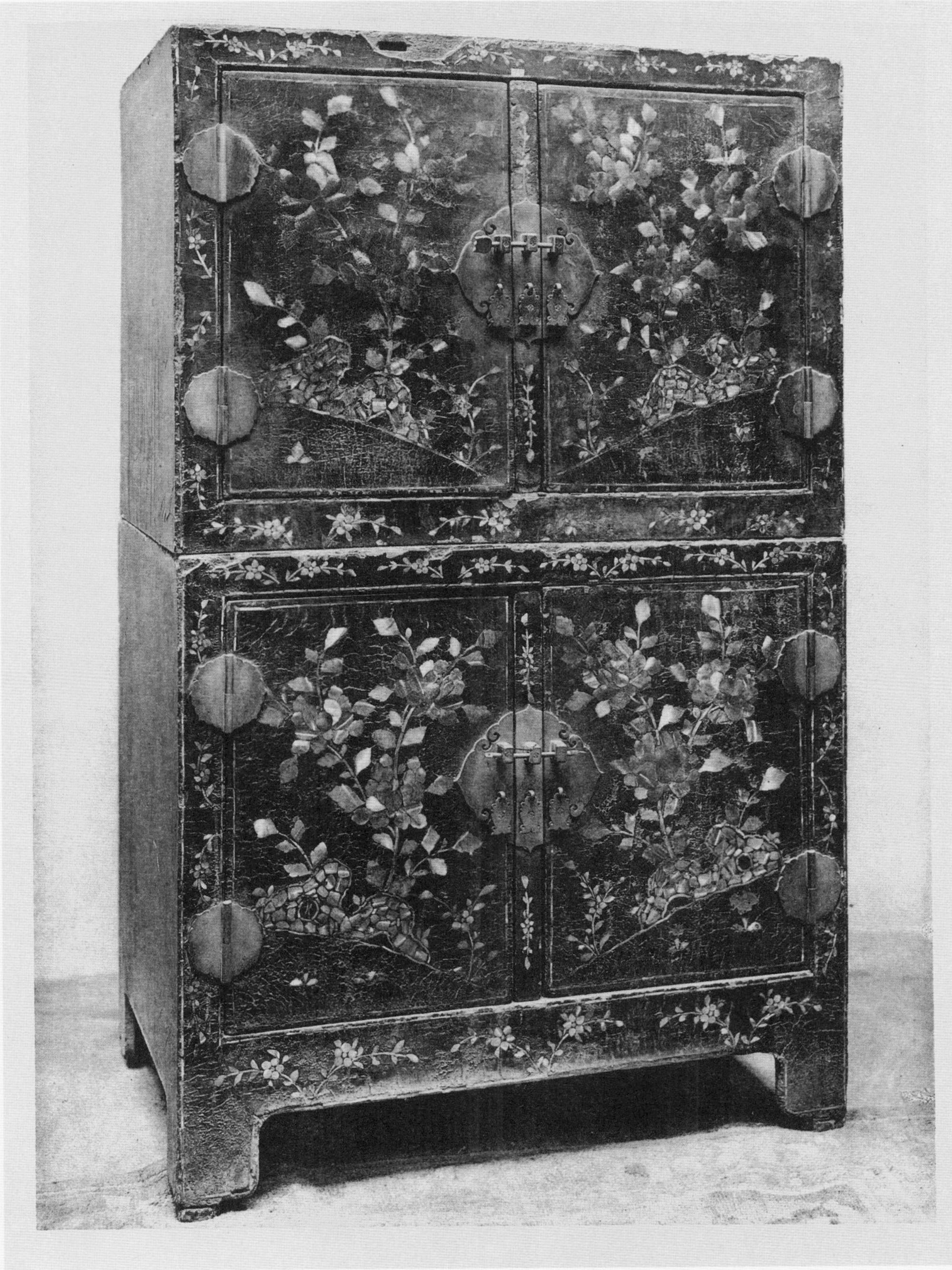

图版 9
PLATE IX.

顶箱柜，黑漆，百宝嵌
高 245.4 厘米　宽 125.7 厘米　深 62.2 厘米

奥迪隆·罗什先生藏

Double-chest, black lacquer, polychrome decoration with mother-of-pearl inlaid and carved. Height, 8 feet $\frac{5}{8}$ inch; breath, 4 feet $1\frac{1}{2}$ inches; depth, 2 feet $\frac{1}{2}$ inch.

In the possession of O. Rothe, Esq.

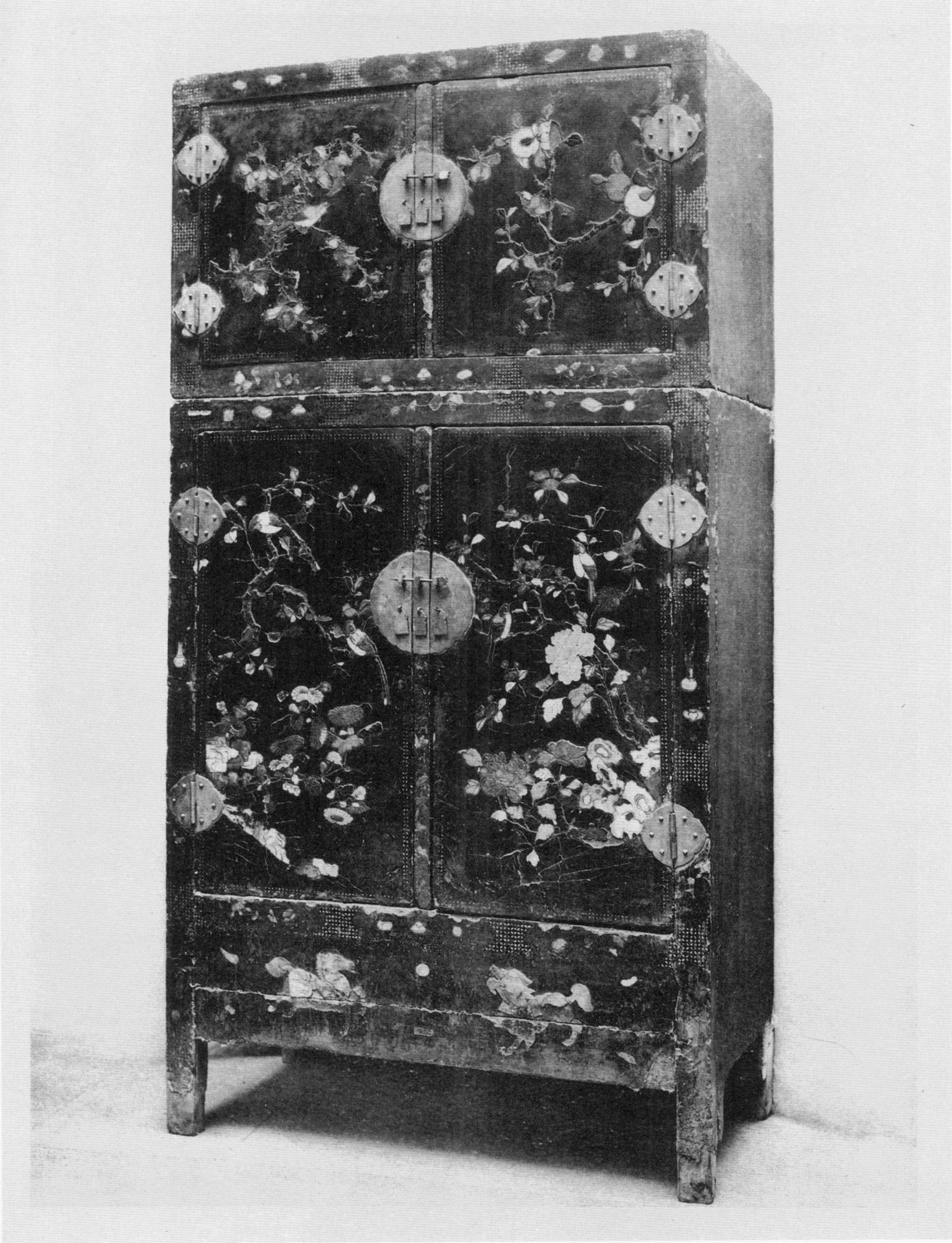

图版 10
PLATE X.

顶箱柜，黑漆，描金
高 311.5 厘米　宽 192.4 厘米　深 82.2 厘米

L. 万尼克先生藏

Double-chest, black lacquer, gold decoration. Height, 10 feet $2\frac{5}{8}$ inches; breadth, 6 feet $3\frac{3}{4}$ inches; depth, 2 feet $8\frac{3}{8}$ inches.

In the possession of L. Wannieck, Esq.

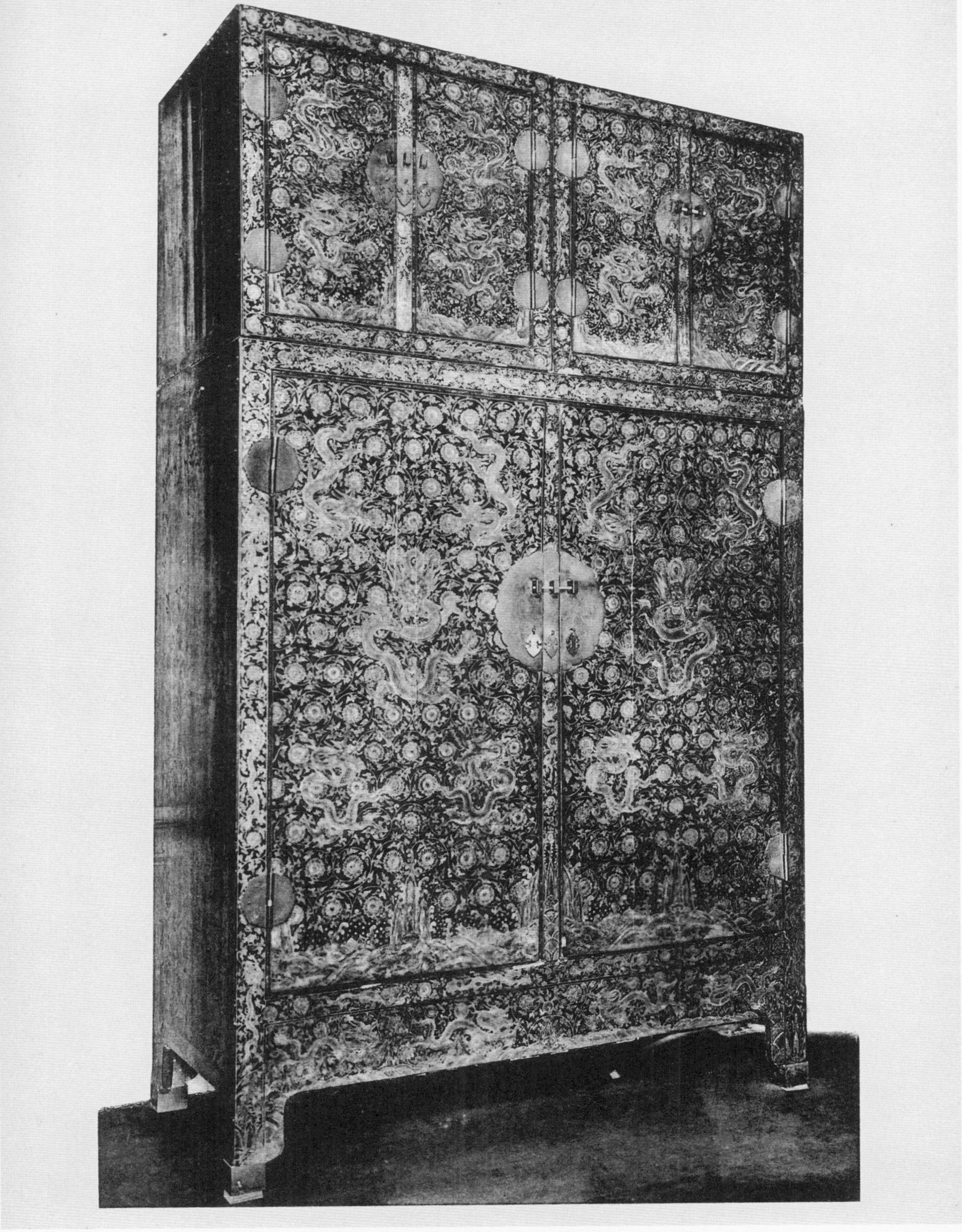

图版 11
PLATE XI.

柜，黑漆，嵌螺钿
高 157.5 厘米　宽 121.6 厘米　深 49.5 厘米

奥迪隆·罗什先生藏

Chest, black lacquer, mother-of-pearl inlaid and carved. Height, 5 feet 2 inches; breadth, 3 feet $11\frac{7}{8}$ inches; depth, 1 foot $7\frac{1}{2}$ inches.

In the possession of O. Roche, Esq.

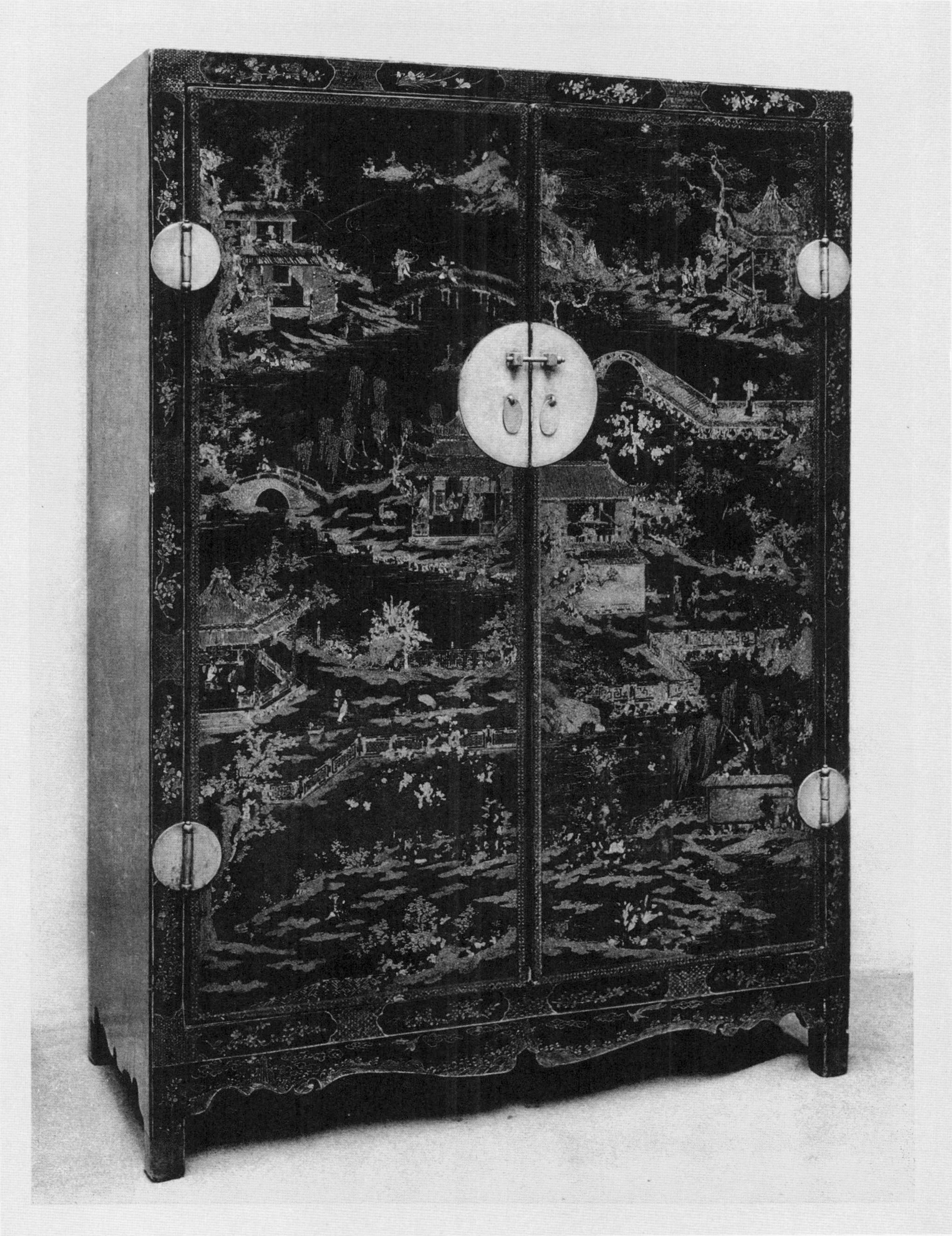

图版 12
PLATE XII.

小橱，黑漆，嵌螺钿
高 139.7 厘米　宽 71.1 厘米　深 36.8 厘米

卢吴公司藏

Small cabinet, black lacquer, mother-of-pearl inlaid and carved. Height, 4 feet 7 inches; breadth, 2 feet 4 inches; depth, 1 foot 2½ inches.

In the possession of Messrs. Loo & Co.

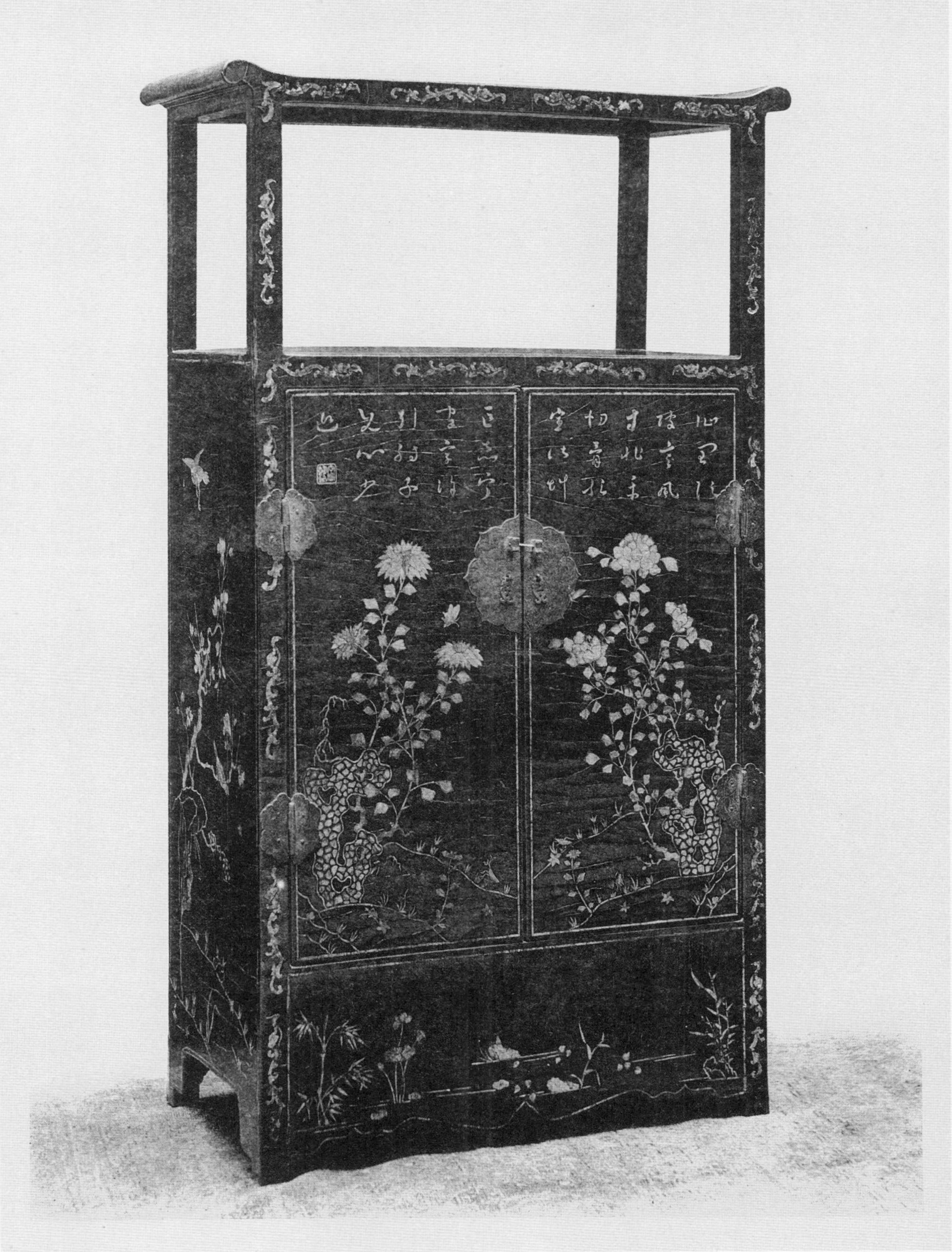

图版 13
PLATE XIII.

顶箱柜，浅黄漆，填彩漆
高 278.8 厘米　宽 144.5 厘米　深 56.5 厘米

奥迪隆·罗什先生藏

Double-chest, buff lacquer, polychrome incised decoration. Height, 9 feet $1\frac{3}{4}$ inches; breadth, 4 feet $8\frac{7}{8}$ inches; depth, 1 foot $10\frac{1}{4}$ inches.

In the possession of O. Roche, Esq.

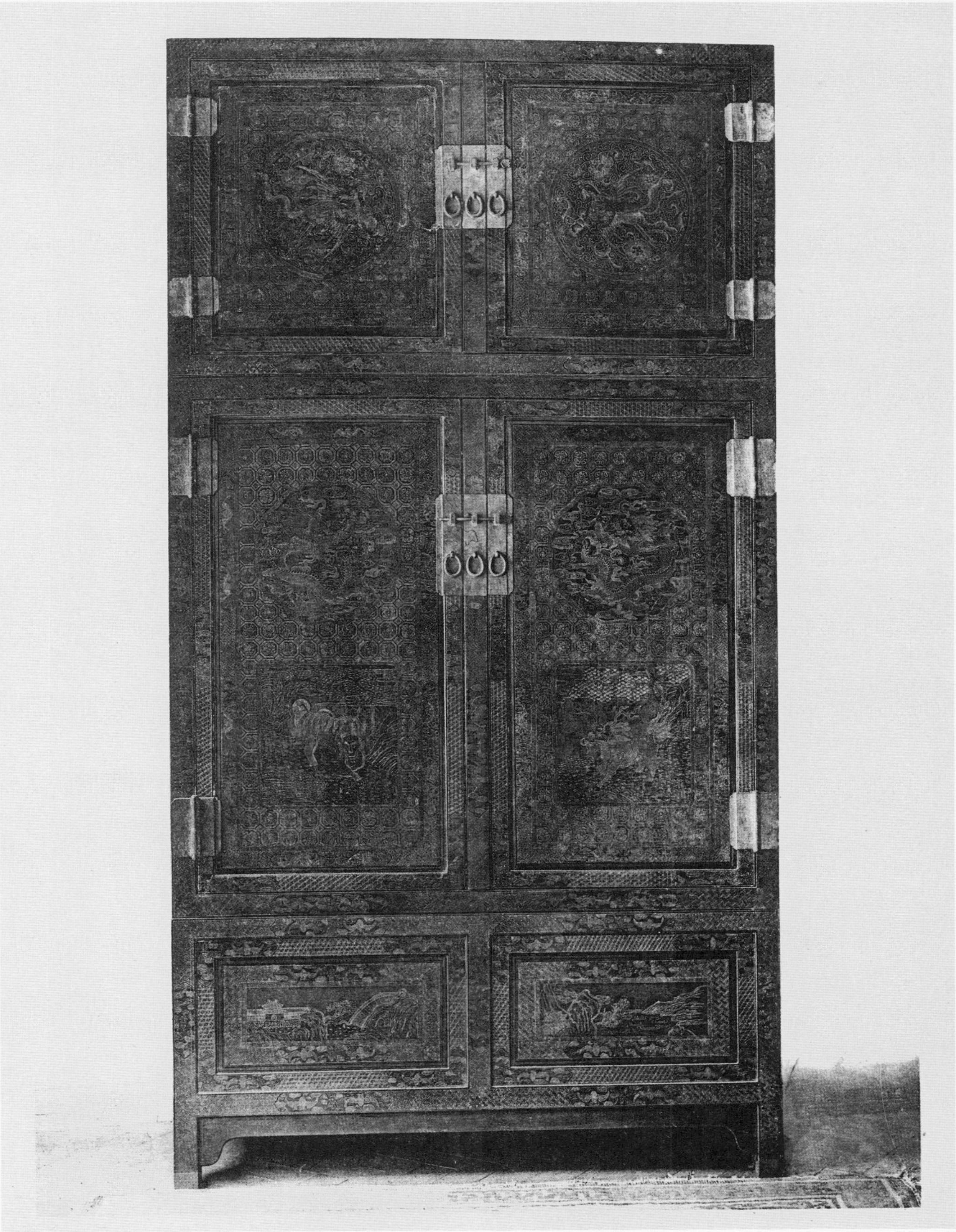

图版 14
PLATE XIV.

柜，褐漆，填彩漆
高 99.7 厘米　宽 149.2 厘米　深 49.5 厘米

奥迪隆·罗什先生藏

Chest, brown lacquer, polychrome incised decoration. Height, 3 feet 3¼ inches; breath, 4 feet 10¾ inches; depth, 1 foot 7½ inches.

In the possession of O. Roche. Esq.

PLATE XIV

箱，褐漆，彩繪裝飾。
高91.5厘米，寬149.2厘米，深33.53厘米。
地點：蕭作氏藏。

Chest, brown lacquer, polychrome
incised decoration. Height, 3 feet
½ inches; breadth, 4 feet 10¾ inches;
depth, 1 foot 1¼ inches.

In the possession of O. Roone, Esq.

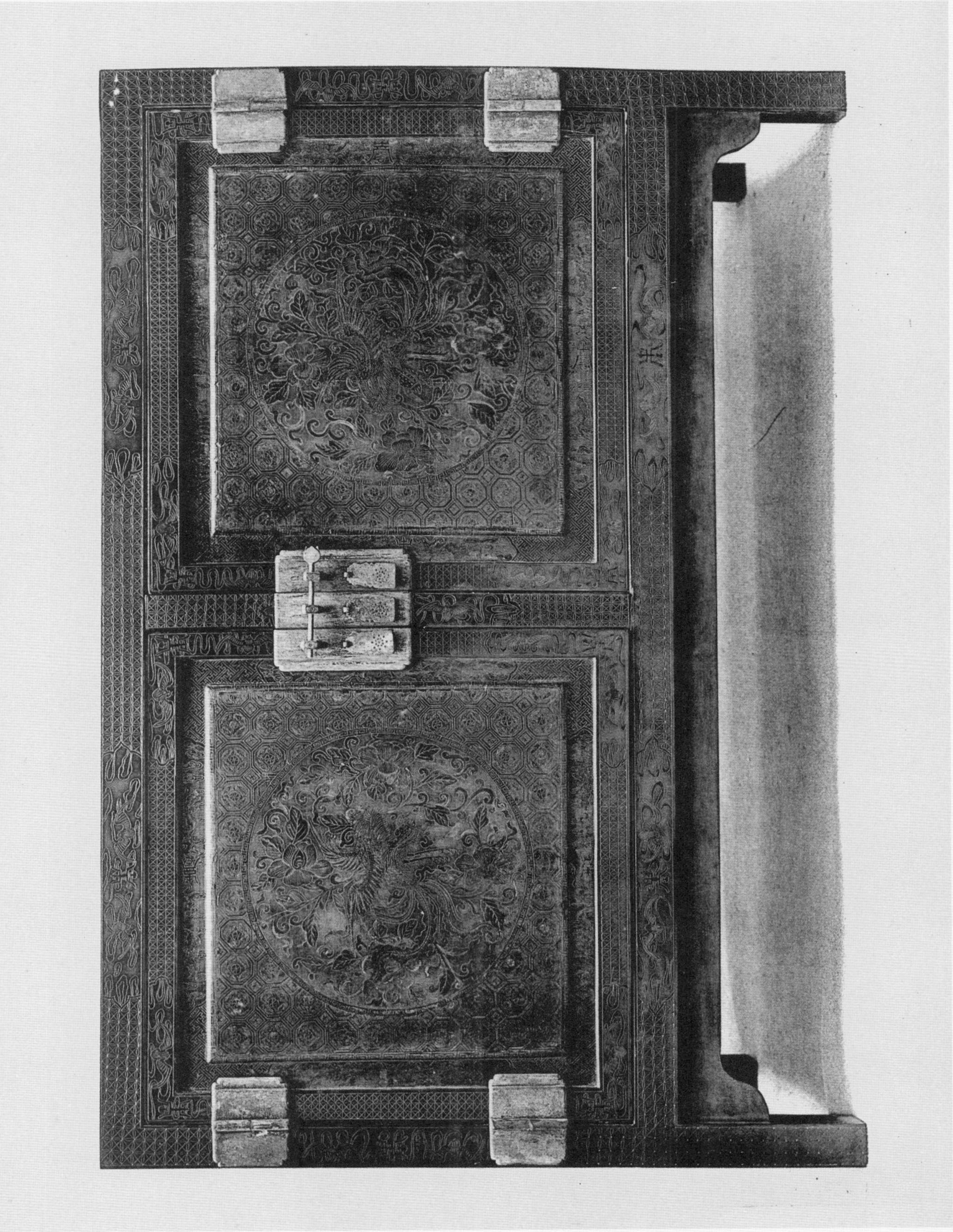

图版 15
PLATE XV.

橱，黑漆，描金
高 142.2 厘米　宽 77.5 厘米　深 47.6 厘米

沃尔希先生藏

Cabinet, black lacquer, gold decoration. Height, 4 feet 8 inches; breadth, 2 feet 6½ inches; depth, 1 foot 6¾ inches.

Worch Colletion.

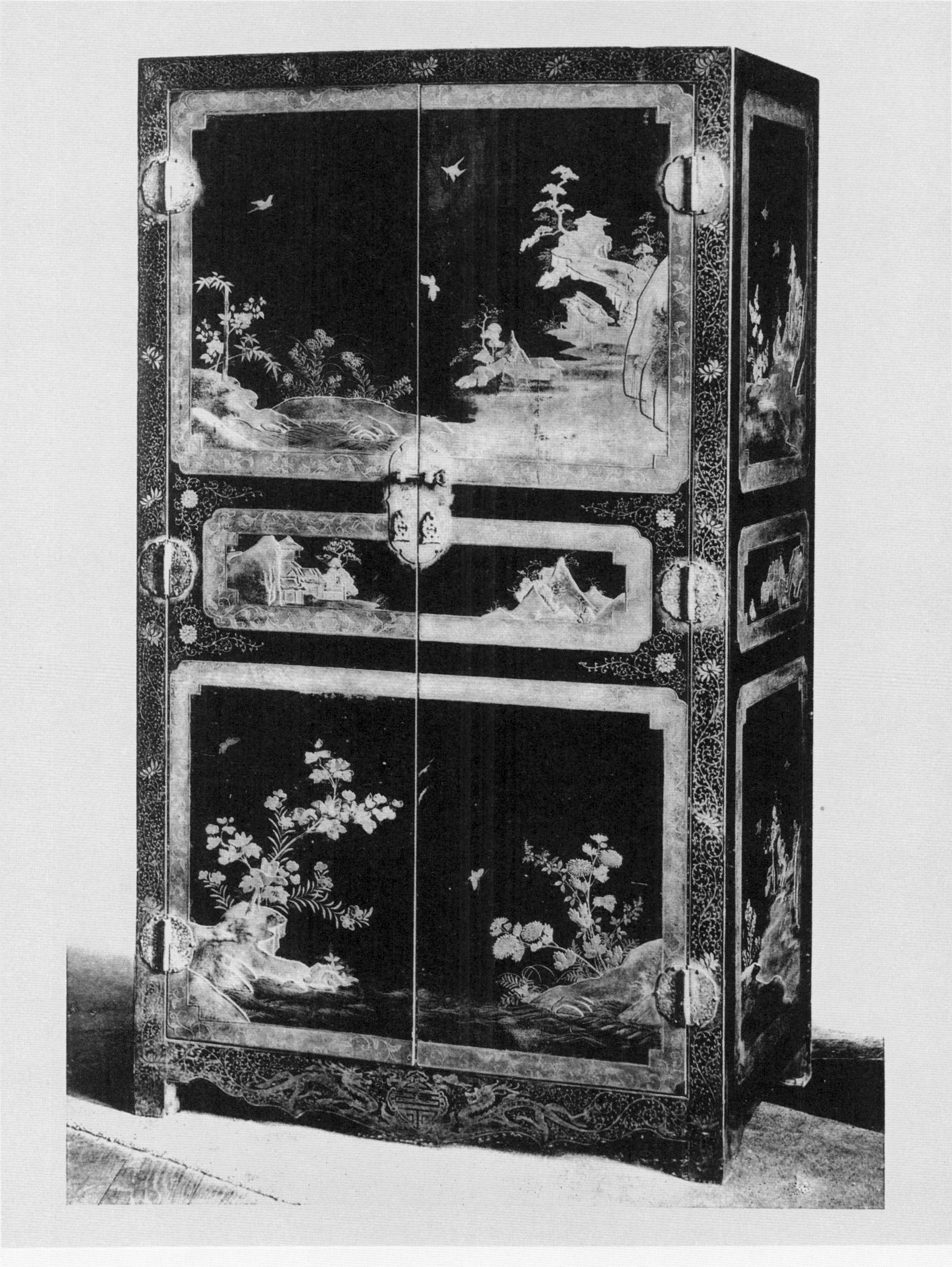

图版 16
PLATE XVI.

顶箱柜,木质,百宝嵌
高 259.1 厘米　宽 141.6 厘米　深 62.2 厘米

L. 万尼克先生藏

Double-chest, natural wood, mother-of-pearl and enamel inlaid and carved. Height, 8 feet 6 inches; breadth, 4 feet 7¾ inches; depth, 2 feet ½ inch.

In the possession of L. Wannieck, Esq.

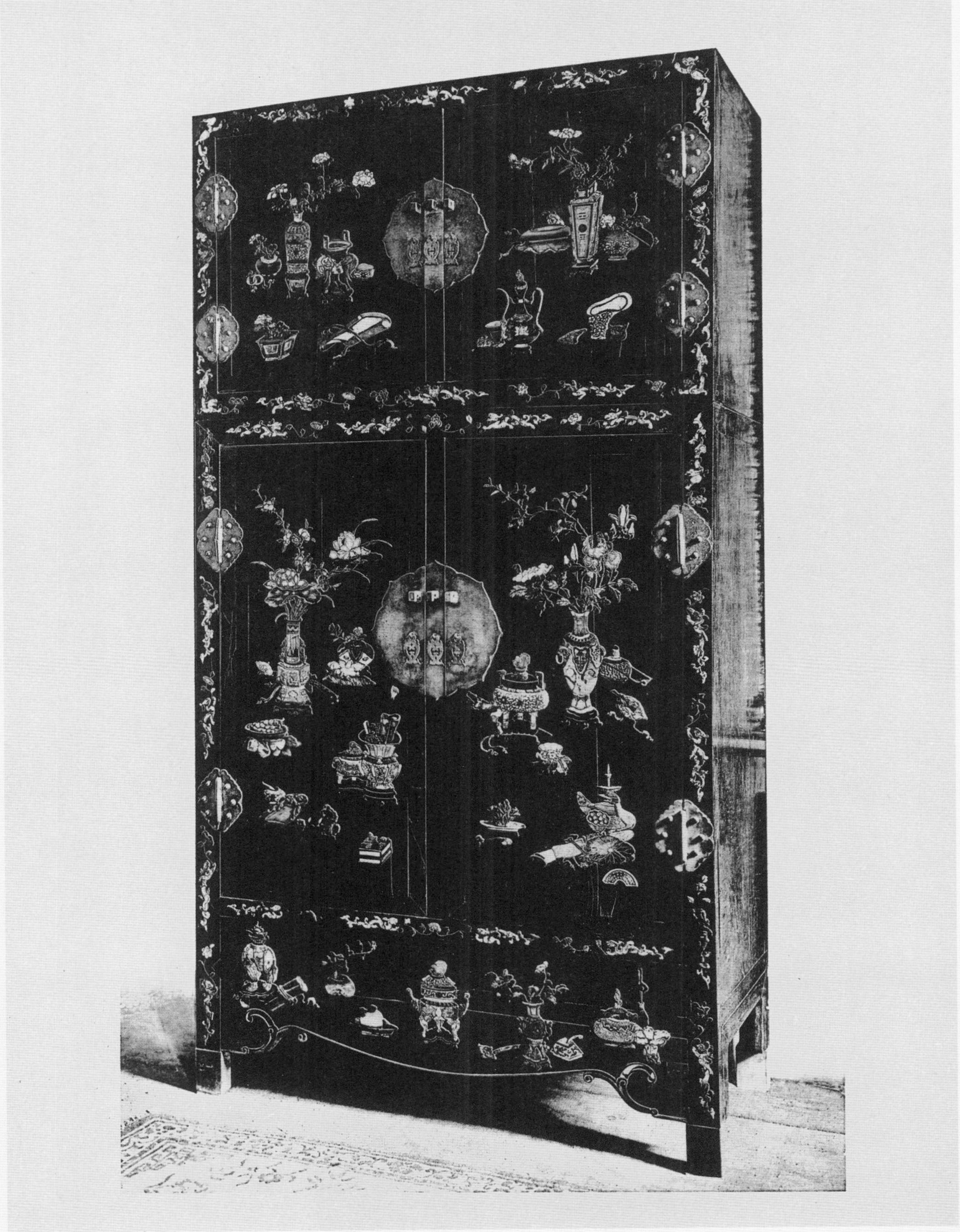

图版 17
PLATE XVII.

柜，红漆，描金
高 204.5 厘米　宽 113.7 厘米　深 53.3 厘米

中国通运公司藏

Chest, red lacquer, gold decoration. Height, 6 feet 8½ inches; breadth, 3 feet 8¾ inches; depth, 1 foot 9 inches.

In the possession of the Chinese Tongying Co.

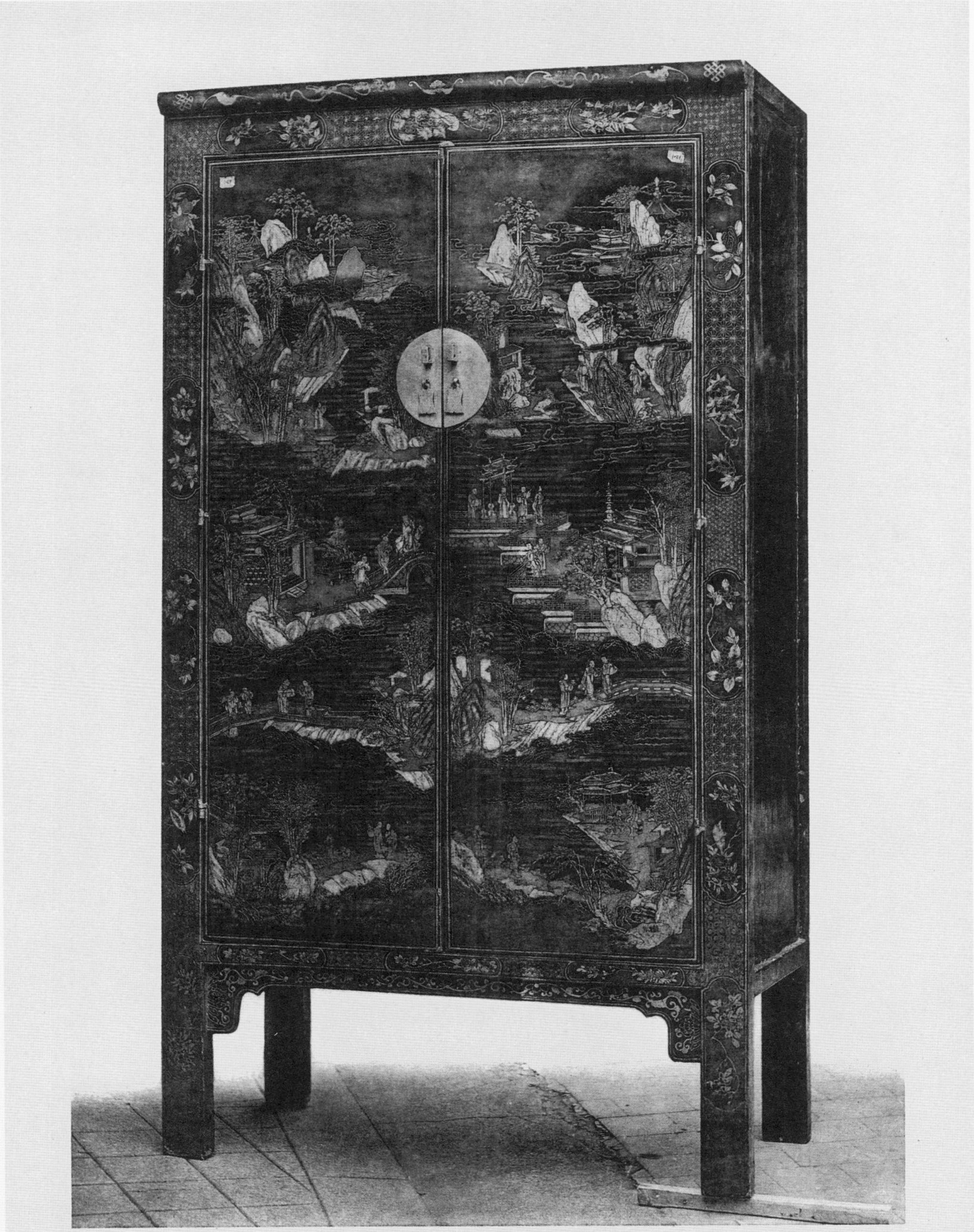

图版 18
PLATE XVIII.

小橱，红漆，描金
高 79.1 厘米　宽 63.5 厘米　深 37.5 厘米

中国通运公司藏

Small cabinet, red lacquer, gold decoration. Height, 2 feet 9⅛ inches; breadth, 2 feet 1 inch; depth, 1 foot 2¾ inches.

In the possession of the Chinese Tongying Co.

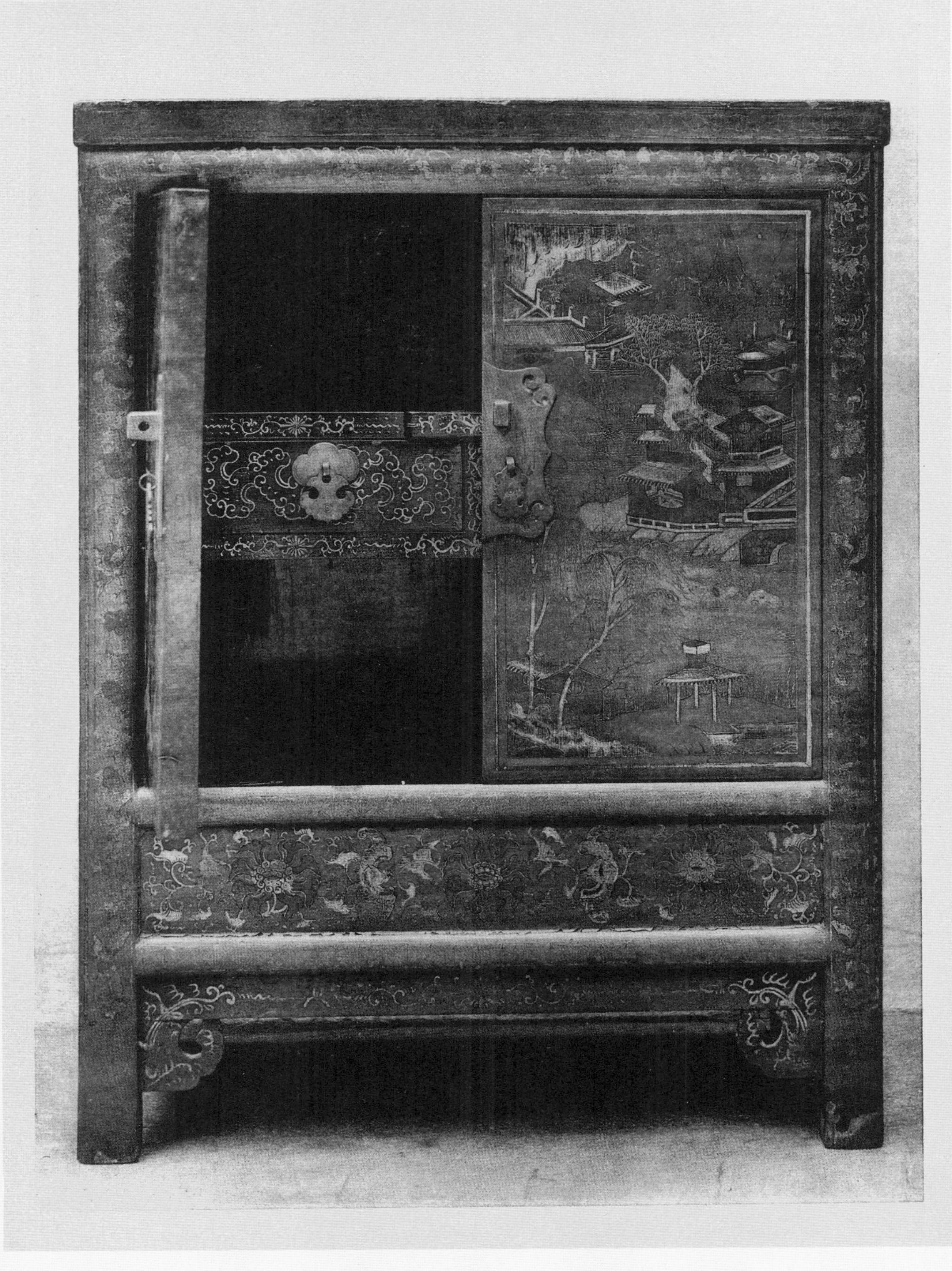

图版 19
PLATE XIX.

柜，黑漆，描红及描金
高 95.9 厘米　宽 189.9 厘米　深 54.6 厘米

中国通运公司藏

Chest, black lacquer, red and gold decoration. Height, 3 feet 1¾ inches; breadth, 6 feet 2¾ inches; depth, 1 foot 9½ inches.

In the possession of the Chinese Tongying Co.

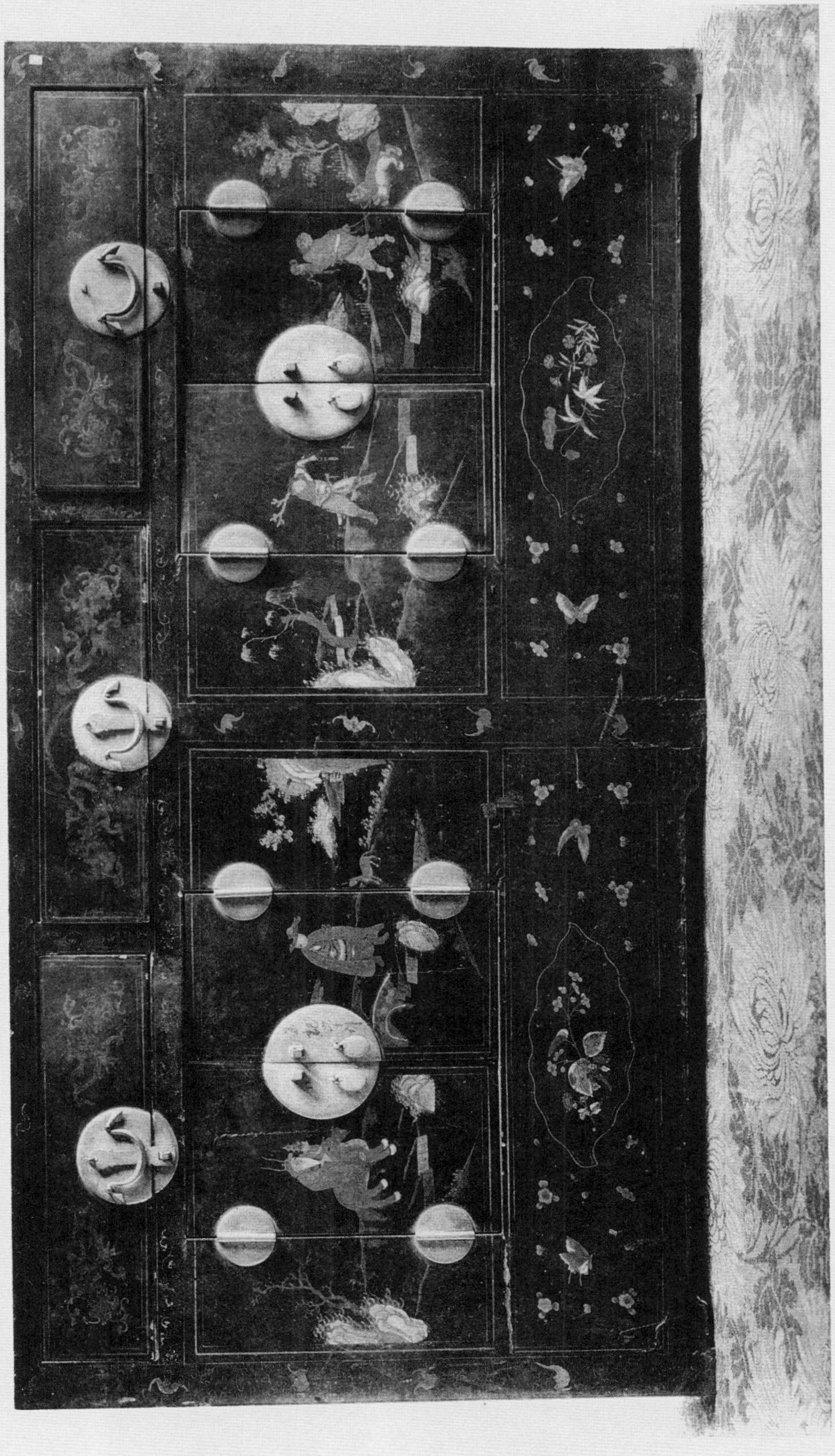

图版 20
PLATE XX.

柜，红漆，描金
高 156.2 厘米　宽 114.9 厘米　深 43.8 厘米

奥迪隆·罗什先生藏

Chest, red lacquer, gold decoration. Height, 5 feet 1½ inches; breadth, 3 feet 9¼ inches; depth, 1 foot 5¼ inches.

In the possession of O. Roche, Esq.

图版 20

PLATE XX.

柜，红漆，描金。
高 56.2 厘米，宽 114.9 厘米，深 54.5 厘米。
现藏者：'ve 奇·罗先生藏

Chest, red lacquer, gold decoration.
Height, 5 feet 14 inches; breadth,
5 feet 4 inches; depth, 1 foot 54
inches.

In the possession of O. Rocke, Esq.

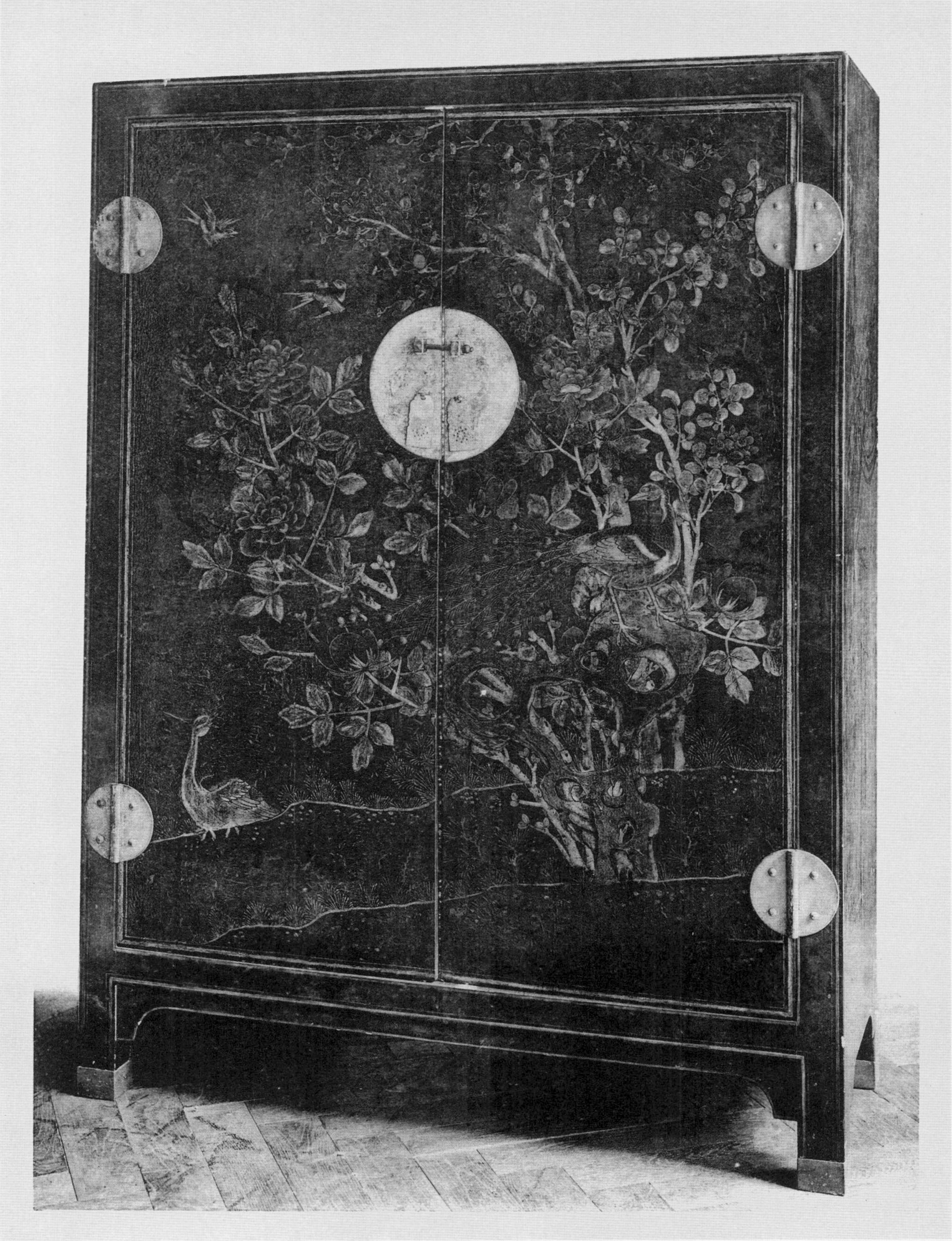

图版 21
PLATE XXI.

柜,红漆,浮雕描金
高 175.9 厘米 宽 147.3 厘米 深 59.4 厘米

查尔斯·维涅先生藏

Chest, red lacquer, gold decoration in relief. Height, 5 feet 9¼ inches; breadth, 4 feet 10 inches; depth, 1 foot 11⅜ inches.

In the possession of Charles Vignier, Esq.

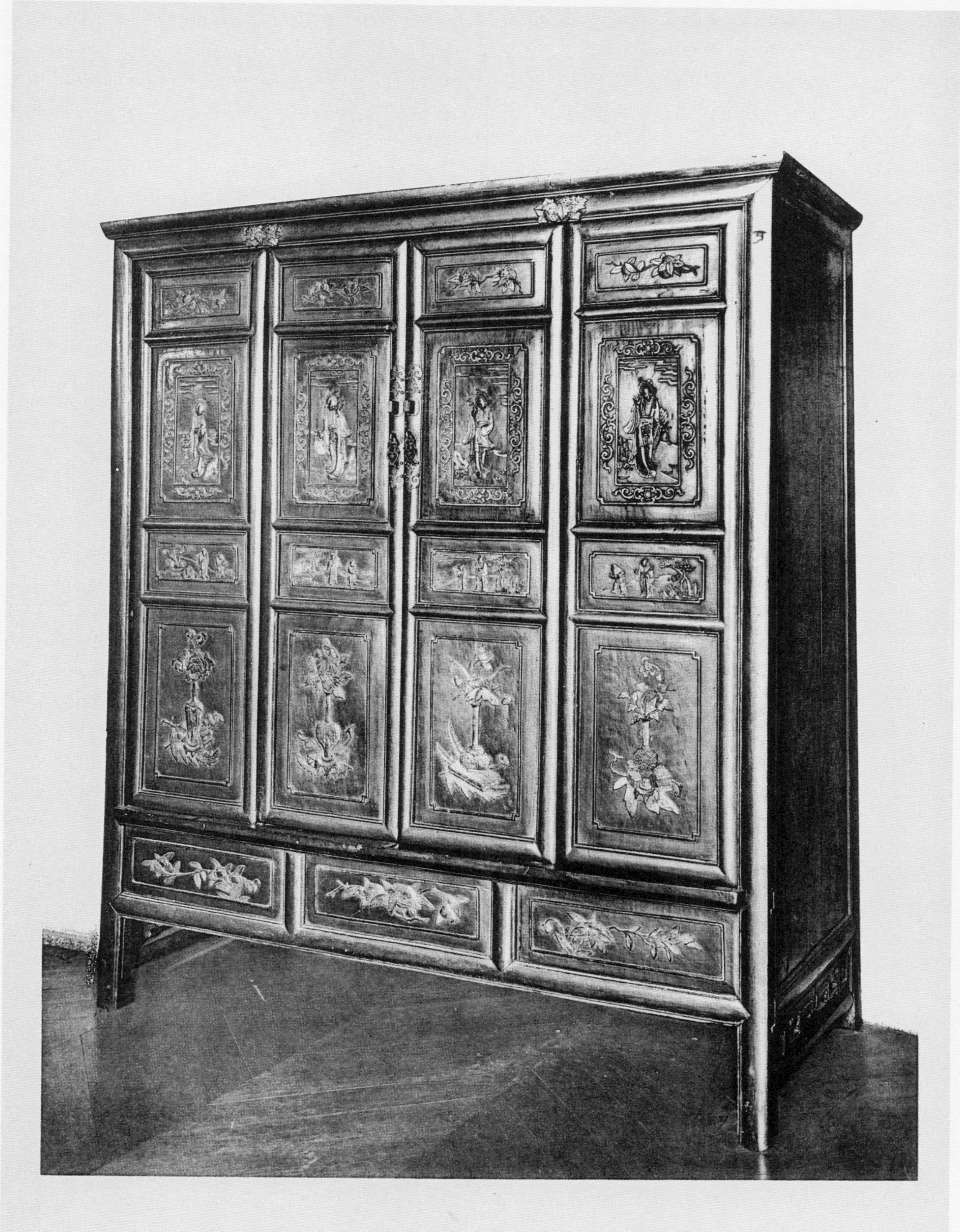

图版 22
PLATE XXII.

(1) 箱，黑漆，嵌螺钿
高 66.4 厘米　宽 89.2 厘米　深 58.4 厘米

奥迪隆·罗什先生藏

(*a*) Coffer, black lacquer, mother-of-pearl inlaid and carved. Height, 2 feet 2$\frac{1}{8}$ inches; breadth, 2 feet 11$\frac{1}{8}$ inches; depth, 1 foot 11 inches.

In the possession of O. Roche, Esq.

(2) 箱，黑漆，彩绘
高 35.6 厘米　宽 89.2 厘米　深 57.5 厘米

保罗·马隆先生藏

(*b*) Coffer, black lacquer, polychrome decoration. Height, 1 foot 2 inches; breadth, 2 feet 11$\frac{1}{8}$ inches; depth, 1 foot 10$\frac{5}{8}$ inches.

In the possession of Paul Mallon, Esq.

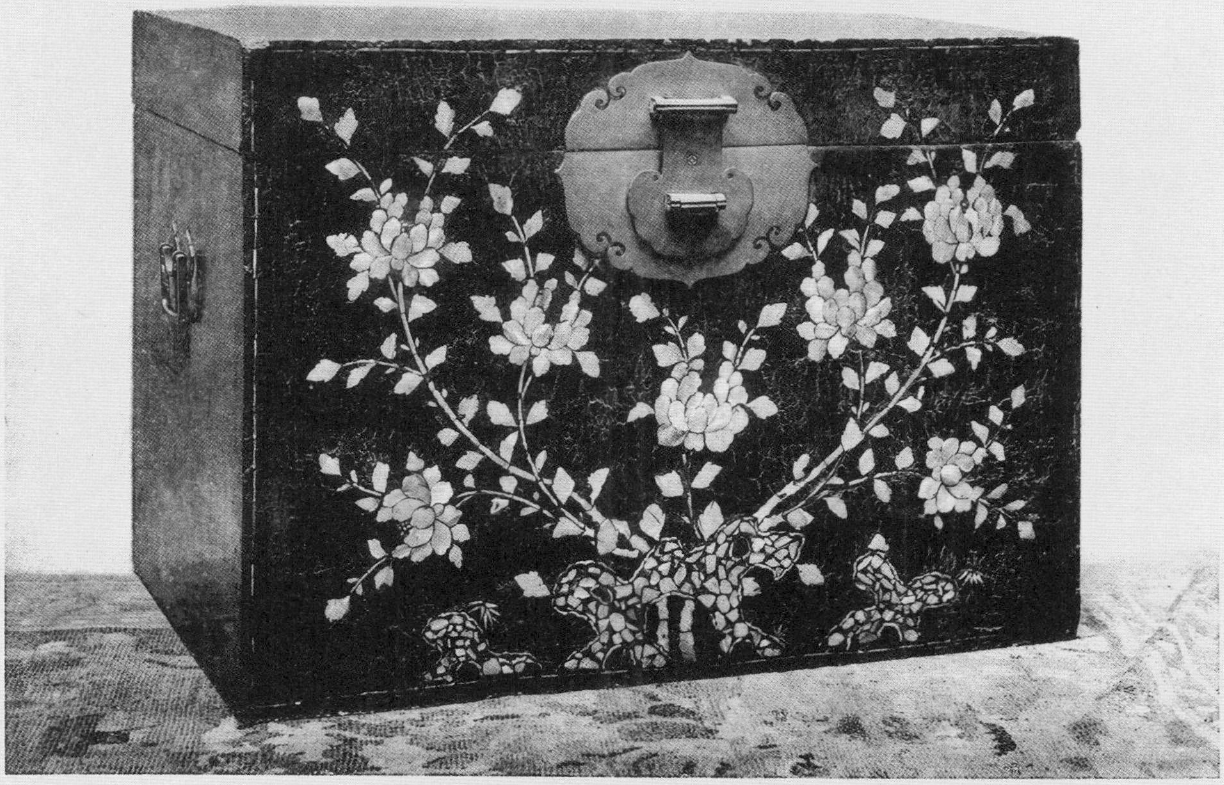
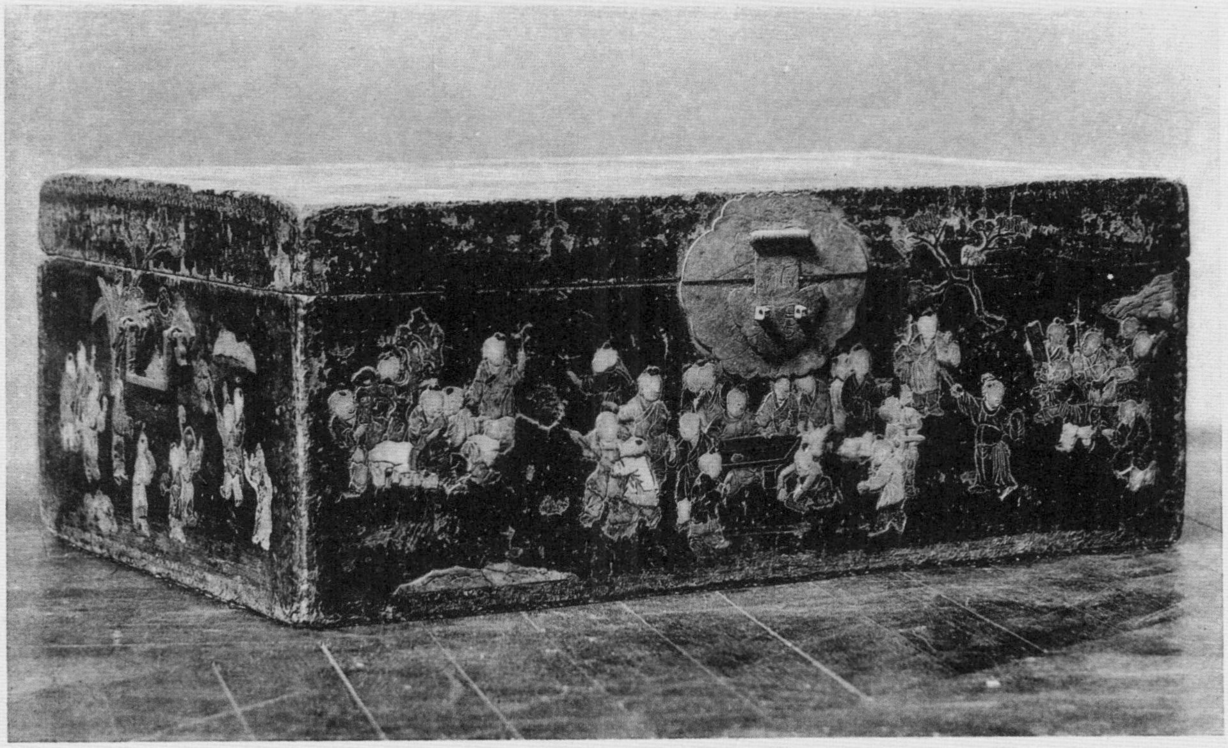

图版 23
PLATE XXIII.

桌，浅黄漆，描红及描黑
长 111.8 厘米　深 82.2 厘米　高 87 厘米

中国通运公司藏

Table, buff lacquer, red and black decoration. Length, 3 feet 8 inches; depth, 2 feet $8\frac{3}{8}$ inches; height, 2 feet $10\frac{1}{4}$ inches.

In the possession of the Chinese Tongying Co.

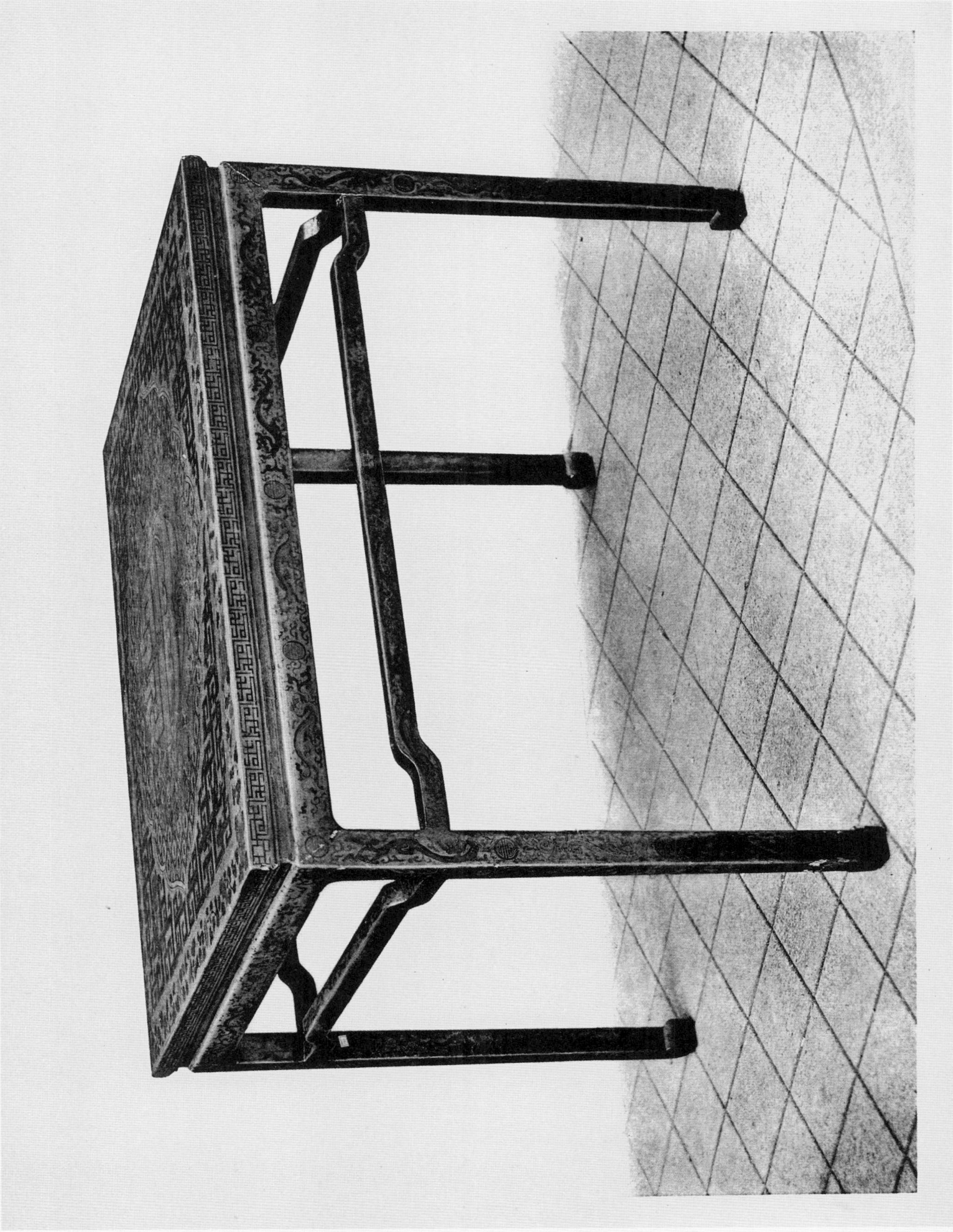

图版 24
PLATE XXIV.

桌，褐漆，彩绘
长 127.6 厘米　深 72.4 厘米　高 84.1 厘米

奥迪隆·罗什先生藏

Table, brown lacquer, polychrome decoration. Length, 4 feet 2¼ inches; depth, 2 feet 4½ inches; height, 2 feet 9⅛ inches.

In the possession of O. Roche, Esq.

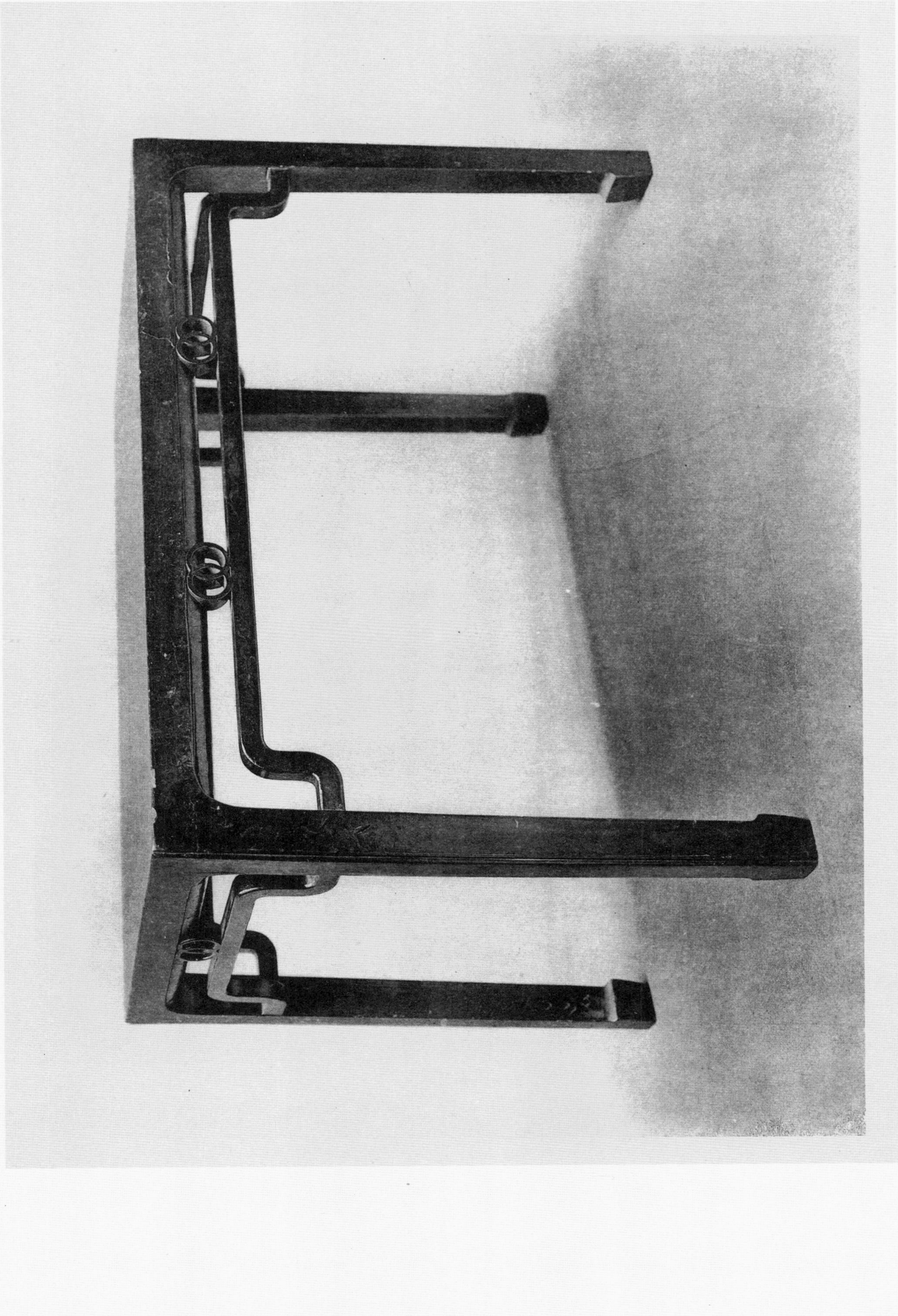

图版 25
PLATE XXV.

桌，浅黄漆，彩绘
长 185.4 厘米　深 54.6 厘米　高 85.1 厘米

奥迪隆·罗什先生藏

Table, buff lacquer, polychrome decoration. Length, 6 feet 1 inch; depth, 1 foot 9½ inches; height, 2 feet 9½ inches.

In the possession of O. Roche, Esq.

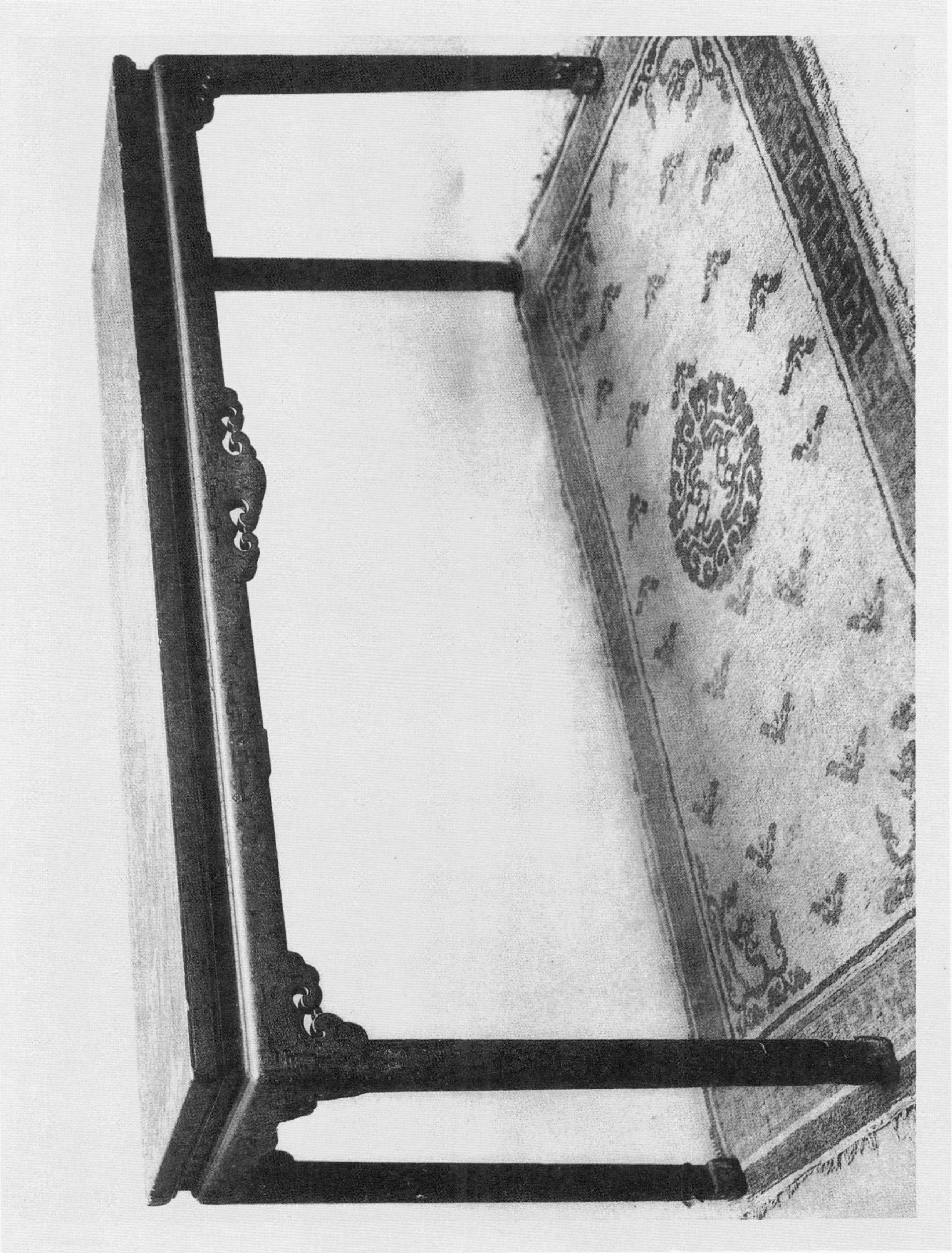

图版 26
PLATE XXVI.

桌，褐漆，嵌螺钿
长 104.8 厘米　深 75.2 厘米　高 81.3 厘米

圣约翰·奥德利先生藏

Table, brown lacquer, mother-of-pearl inlaid and carved. Length, 3 feet $5\frac{1}{4}$ inches; depth, 2 feet $5\frac{5}{8}$ inches; height, 2 feet 8 inches.

In the possession of St. John Audley, Esq.

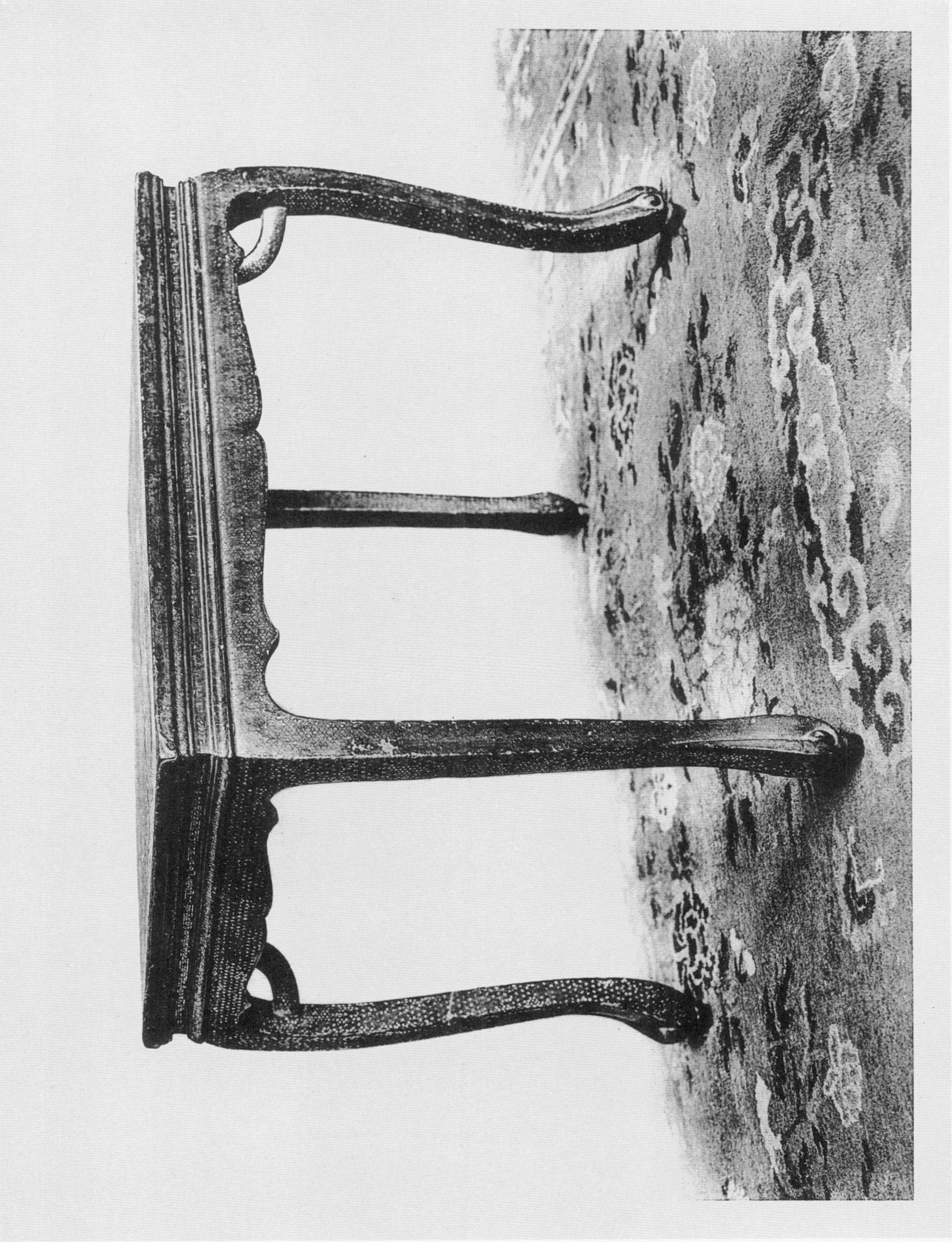

图版 27
PLATE XXVII.

扶手椅和桌，黑漆，嵌螺钿
扶手椅：高 109.9 厘米　宽 59.4 厘米　深 47.6 厘米
桌：高 86.4 厘米　径 124.8 厘米

L. 万尼克先生藏

Arm-chair and Table, black lacquer, mother-of-pearl inlaid and carved. Height of the arm-chair, 3 feet $7\frac{1}{4}$ inches; breadth, 1 foot $11\frac{3}{8}$ inches; depth, 1 foot $6\frac{3}{4}$ inches. Height of the table, 2 feet 10 inches; diameter, 4 feet $1\frac{1}{8}$ inches.

In the possession of L. Wannieck, Esq.

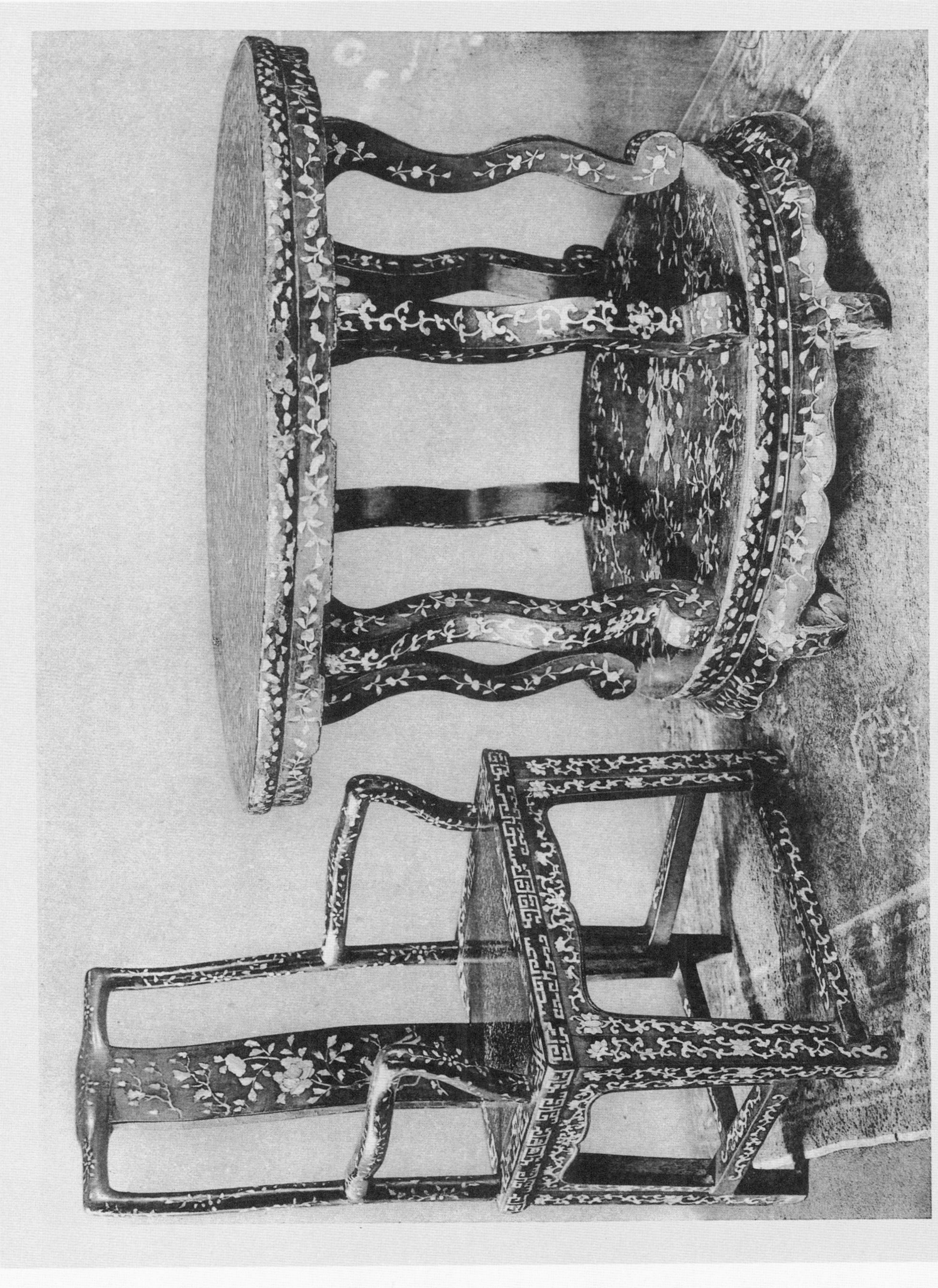

图版 28
PLATE XXVIII.

(1) 方桌，浅黄漆，填彩漆
 长 95.9 厘米　深 95.9 厘米　高 85 厘米

 奥迪隆・罗什先生藏

(a) Square table, buff lacquer, polychrome incised decoration. Length and depth, 3 feet 1¾ inches; height, 2 feet 9½ inches.

 In the possession of O. Roche, Esq.

(2) 桌，素黑漆
 长 74.3 厘米　深 56.5 厘米　高 81.3 厘米

 奥迪隆・罗什先生藏

(b) Table, black lacquer, without ornamentation. Length, 2 feet 5¼ inches; depth, 1 foot 10¼ inches; height, 2 feet 8 inches.

 In the possession of O. Roche, Esq.

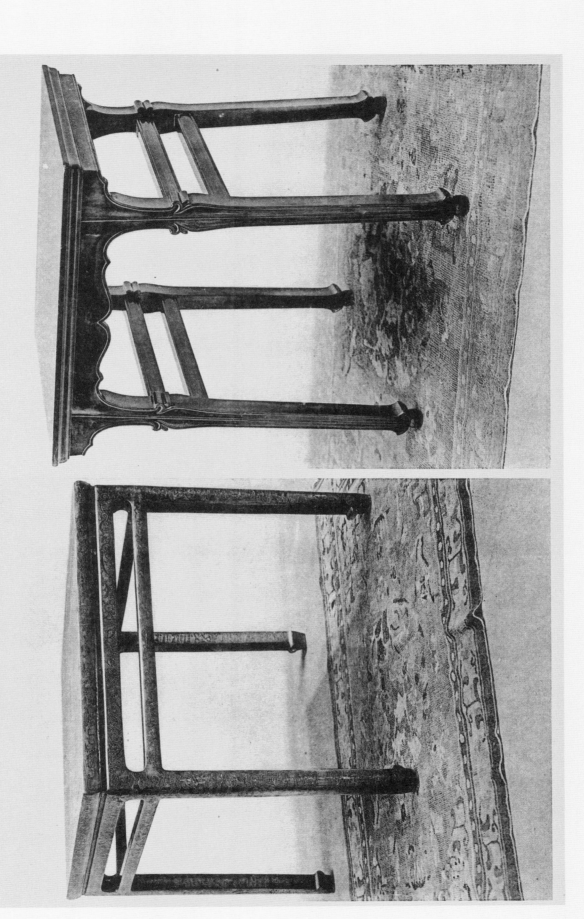

图版 29
PLATE XXIX.

方桌，黑漆，嵌螺钿
长 92.1 厘米　宽 92.1 厘米　高 87 厘米

奥迪隆·罗什先生藏

Square table, black lacquer, mother-of-peal inlaid and carved. Length and depth, 3 feet $\frac{1}{4}$ inch; height, 2 feet 10 $\frac{1}{4}$ inches.

In the possession of O. Roche, Esq.

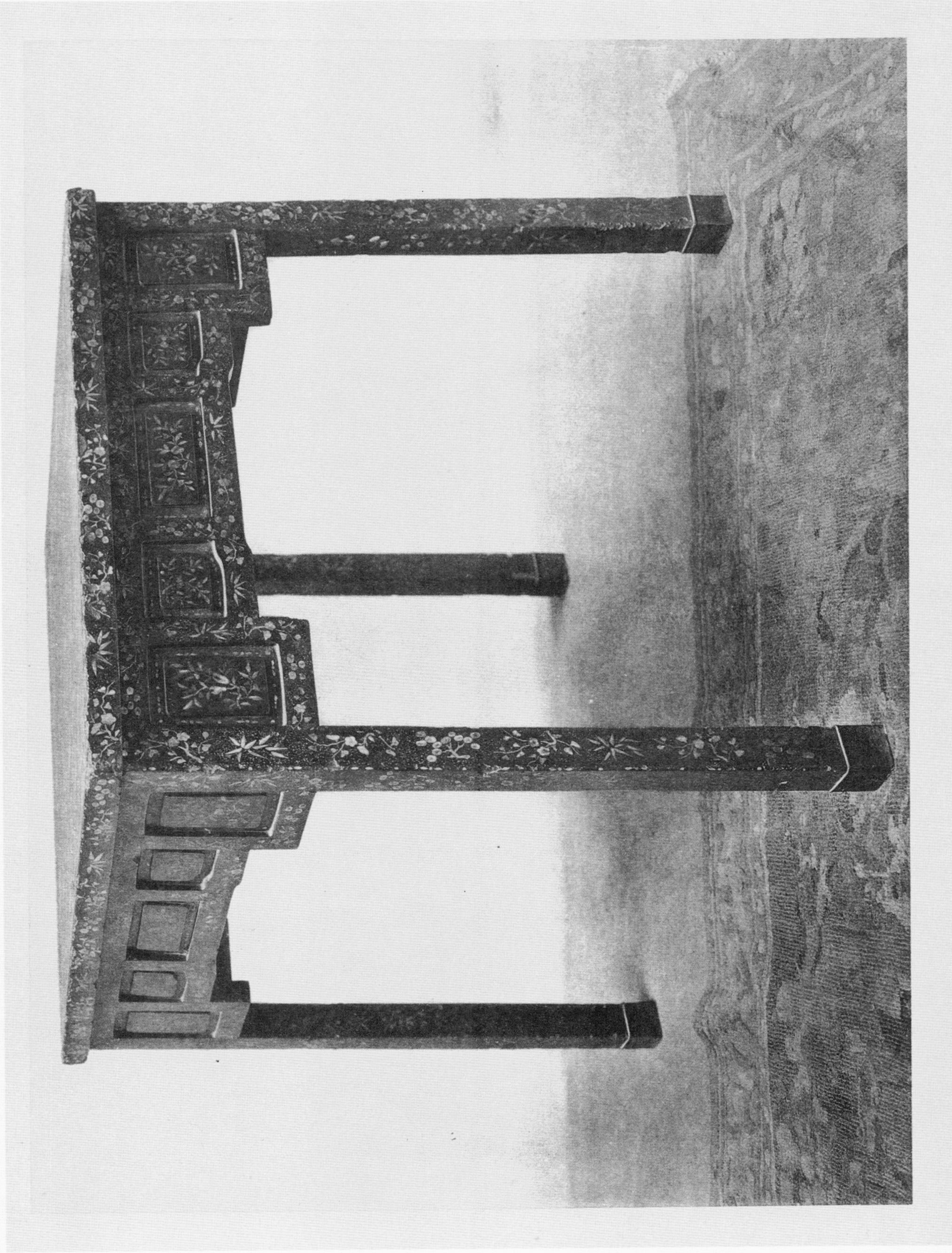

图版 30
PLATE XXX.

桌，浅黄漆，彩绘
长 109.9 厘米　深 72.4 厘米　高 83.2 厘米

保罗·马隆先生藏

Table, buff lacquer, polychrome decoration. Length, 3 feet 7¼ inches; depth, 2 feet 4½ inches; height, 2 feet 8¾ inches.

In the possession of Paul Mallon, Esq.

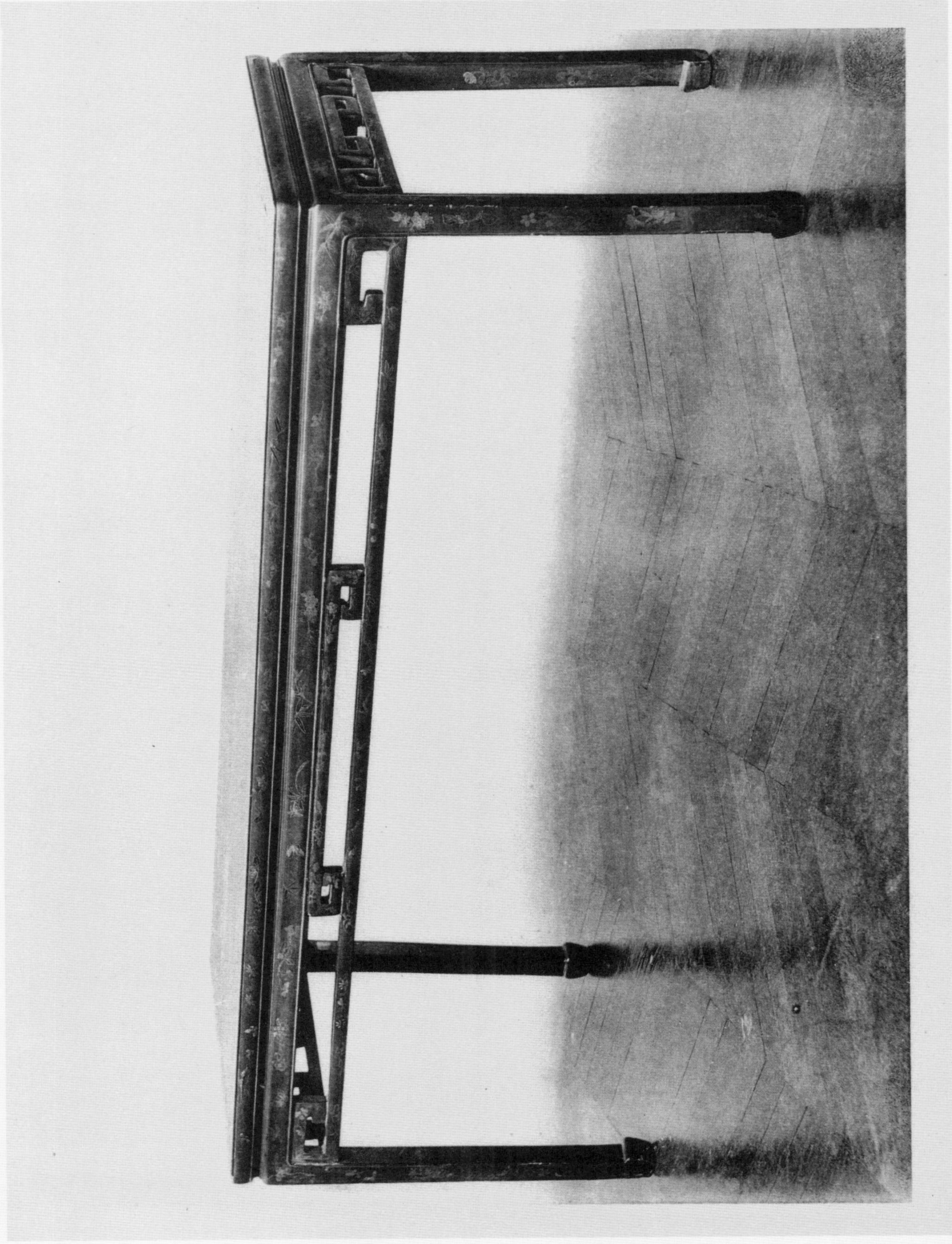

图版 31
PLATE XXXI.

桌，浅黄漆，彩绘
长 215.6 厘米　深 152.4 厘米　高 89.2 厘米

卢吴公司藏

Table, buff lacquer, polychrome decoration. Length, 7 feet $\frac{7}{8}$ inch; depth, 5 feet; height, 2 feet $11\frac{1}{8}$ inches.

In the possession of Mossrs. Loo & Co.

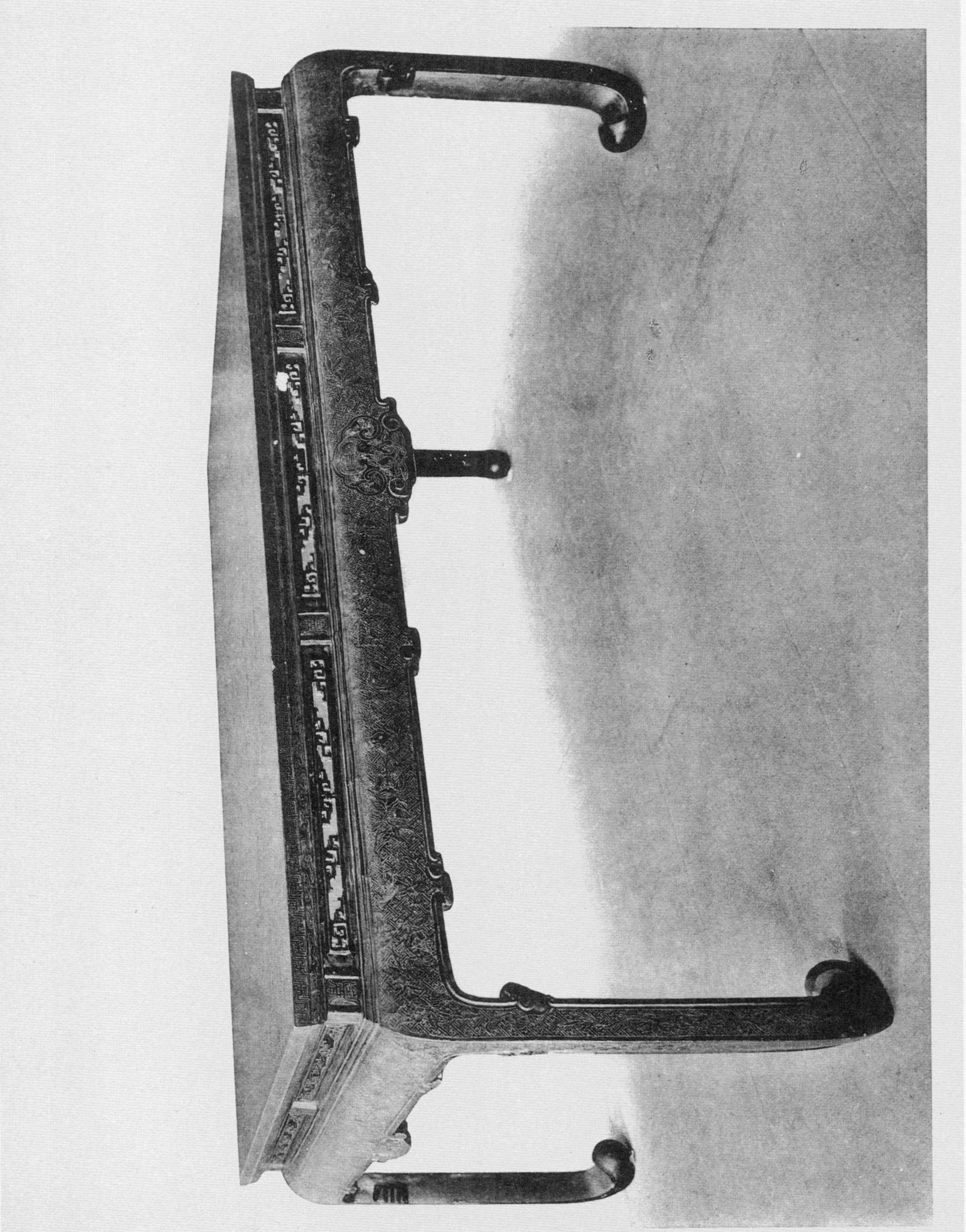

图版 32
PLATE XXXII.

桌，浅黄漆，彩绘
长 114.9 厘米　深 56.5 厘米　高 80 厘米

奥迪隆·罗什先生藏

Table, buffer lacquer, polychrome decoration. Length, 3 feet 9¼ inches; depth, 1 foot 10¼ inches; height, 2 feet 7½ inches.

In the possession of O. Roche, Esq.

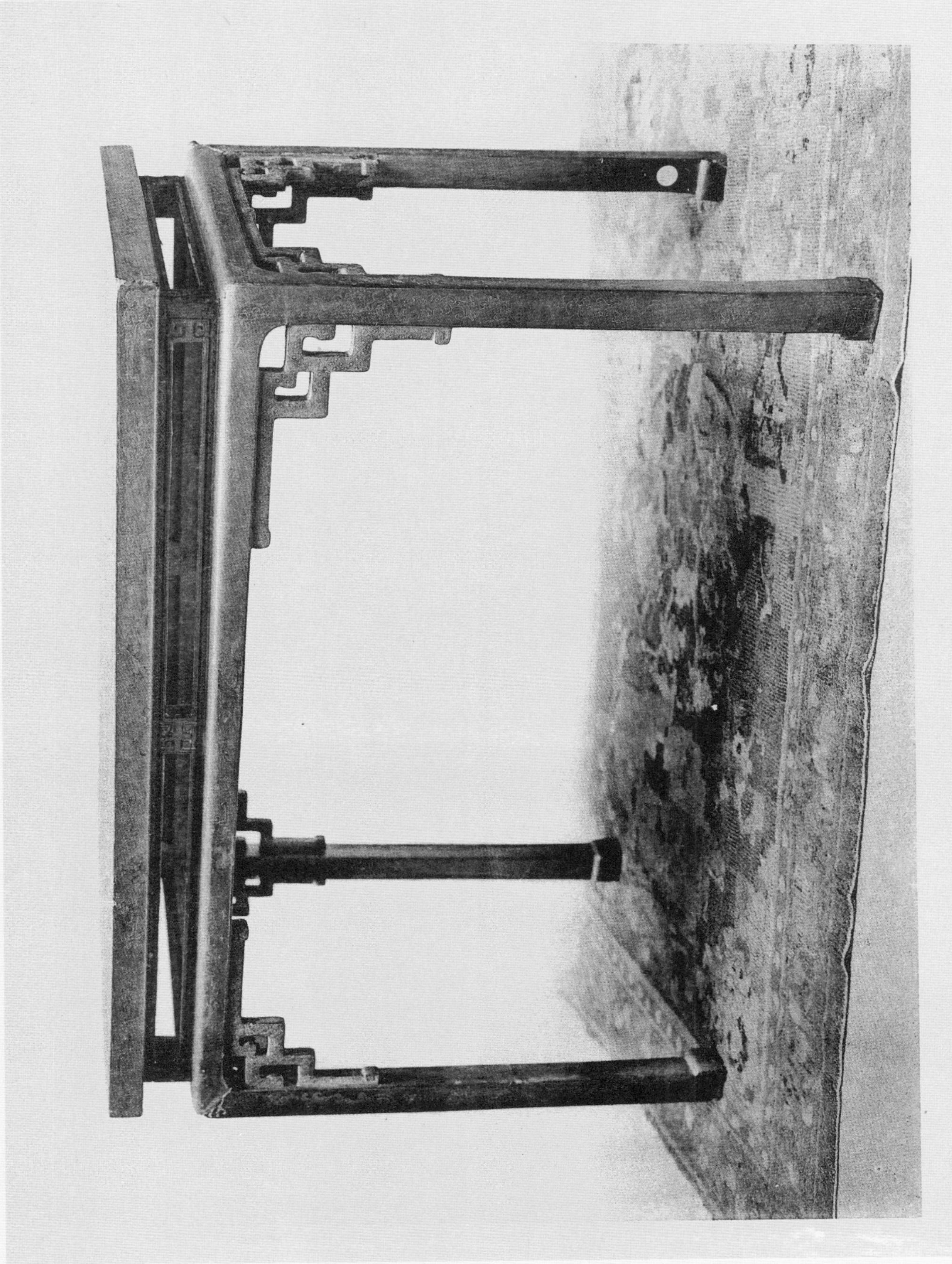

图版 33
PLATE XXXIII.

案，浅黄漆，彩绘
长 257.2 厘米　深 61.6 厘米　高 86.4 厘米

奥迪隆·罗什先生藏

Table, buff lacquer, polychrome decoration. Length, 8 feet 5¼ inches; depth, 2 feet ¼ inch; height, 2 feet 10 inches.

In the possession O. Roche, Esq

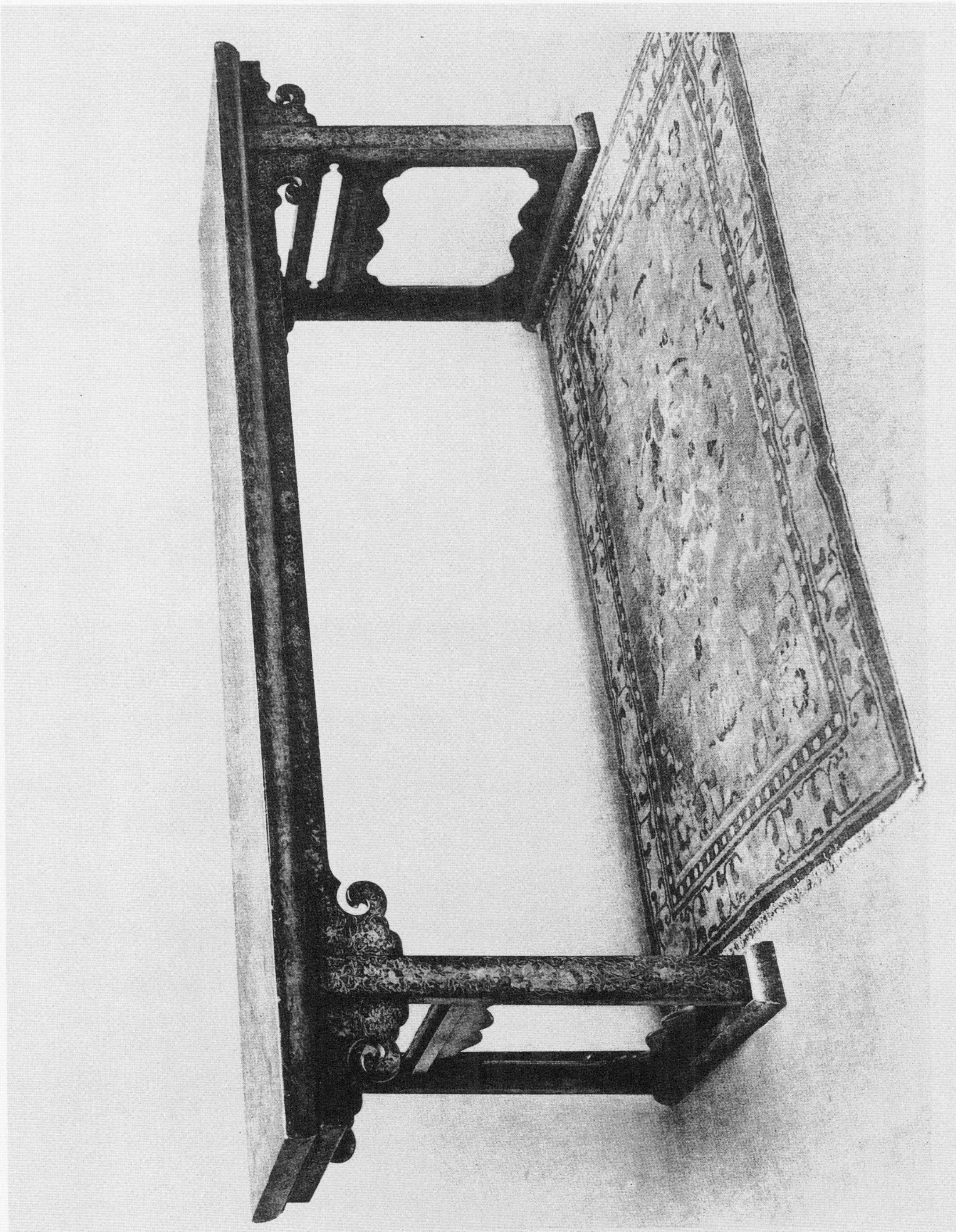

图版 34
PLATE XXXIV.

案（图版 33）的局部
Detail of the table, Plate XXXIII.

图版三
PLATE XXXIV

图 (图版33) 的细部
Detail of the table, Plate XXXIII.

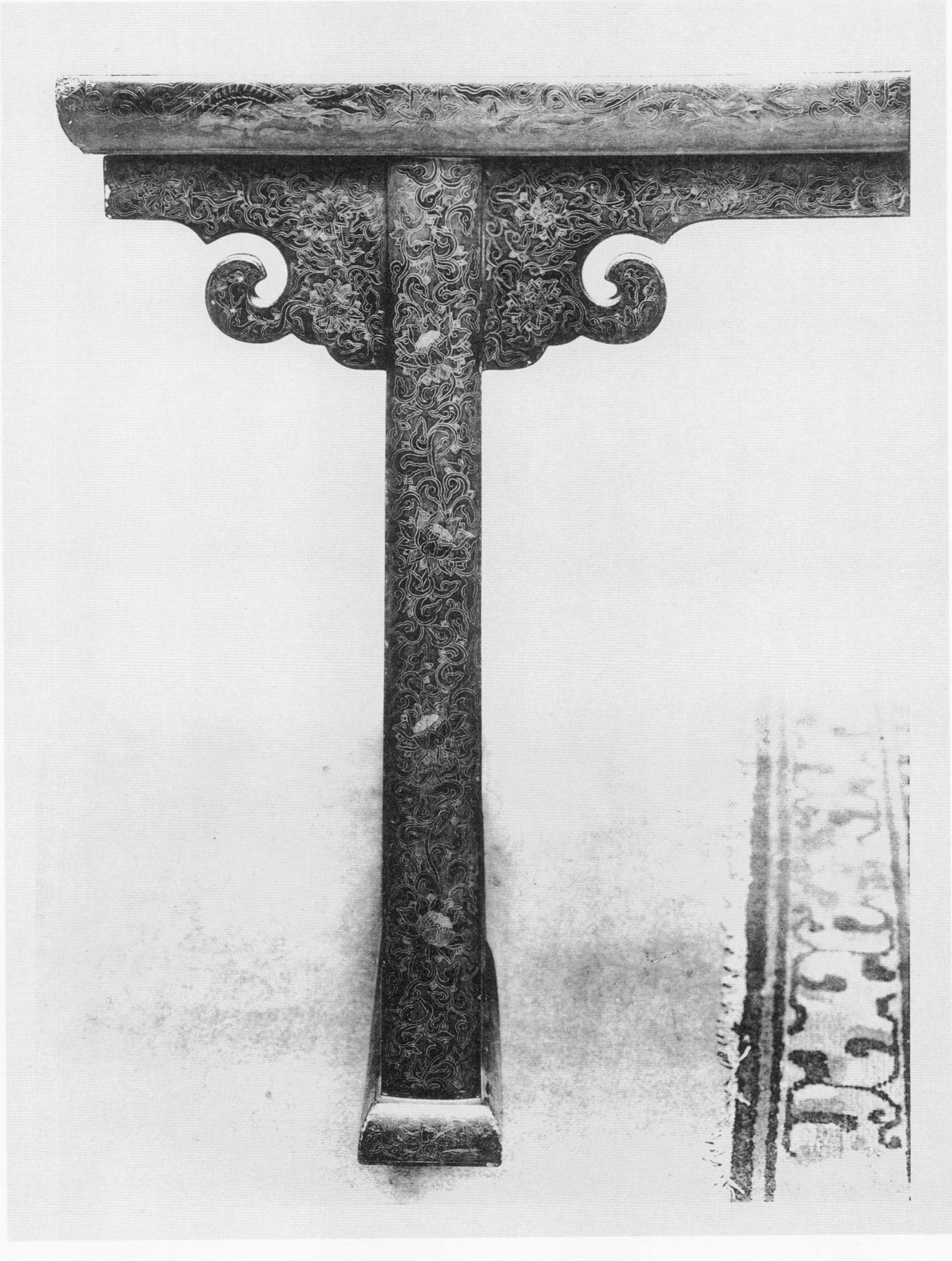

图版 35
PLATE XXXV.

案，浅黄漆，彩绘
长 311.5 厘米　深 66.4 厘米　高 95.9 厘米

圣约翰·奥德利先生藏

Table, buff lacquer, polychrome decoration. Length, 10 feet $2\frac{5}{8}$ inches; depth, 2 feet $2\frac{1}{8}$ inches; height, 3 feet $1\frac{3}{4}$ inches.

In the possession of St. John Audley, Esq.

圖版 35
PLATE XXXV

名: 條凡檯, 彩漆。
長 31.5 厘米, 寬 66 厘米, 高 95.5 厘米
聖約翰 · 奧特利先生藏。

Table, buff lacquer, polychrome
decoration. Length, 10 feet 2½
inches; depth, 2 feet 2½ inches;
height, 3 feet 1½ inches.

In the possession of St. John Audley, Esq.

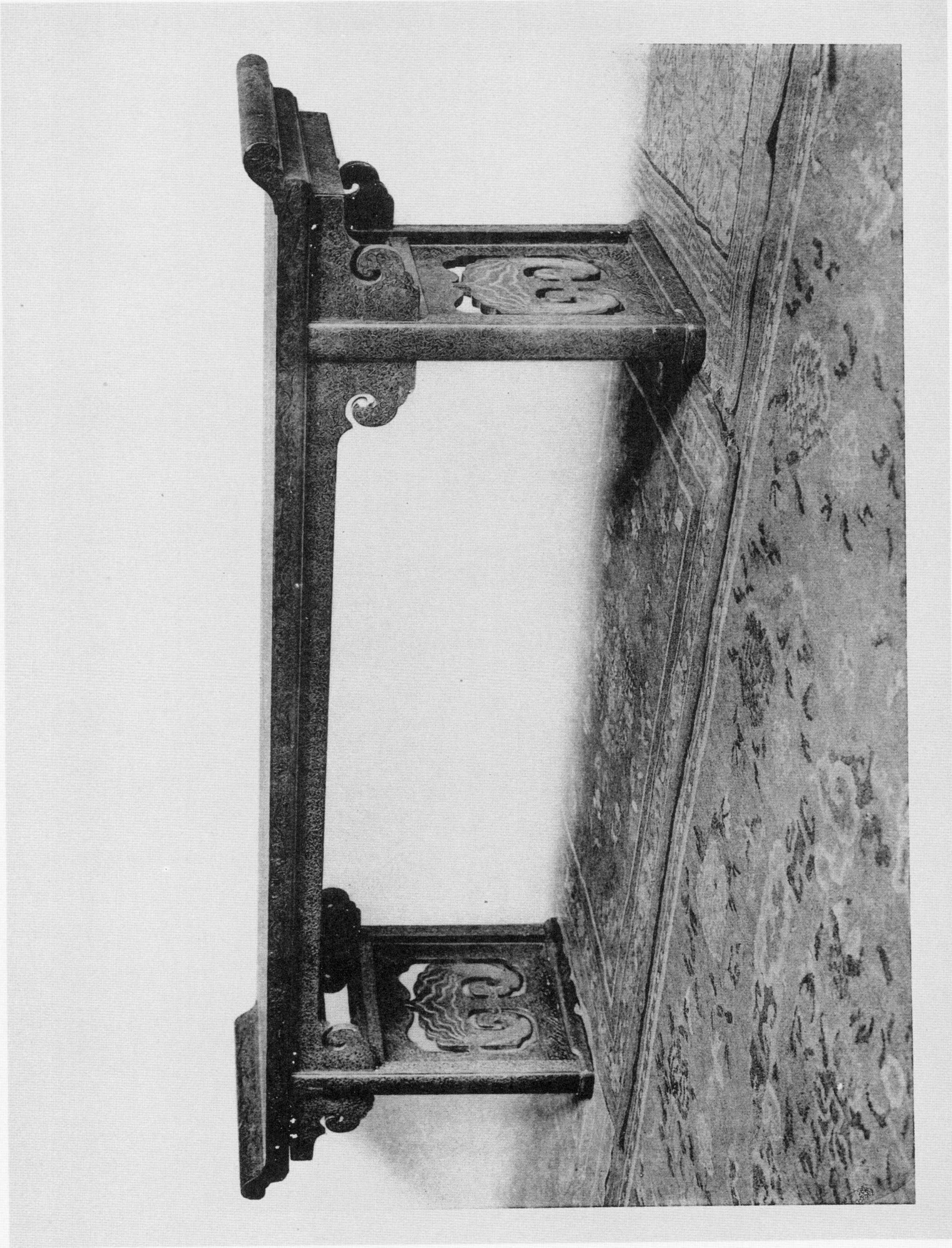

图版 36
PLATE XXXVI.

桌，褐漆，描金
长 119.7 厘米　深 37.5 厘米　高 85.1 厘米

沃尔希先生藏

Table, brown lacquer, gold decoration. Length, 3 feet $11\frac{1}{8}$ inches; depth, 1 foot $2\frac{3}{4}$ inches; height, 2 feet $9\frac{1}{2}$ inches.

Worch Collection.

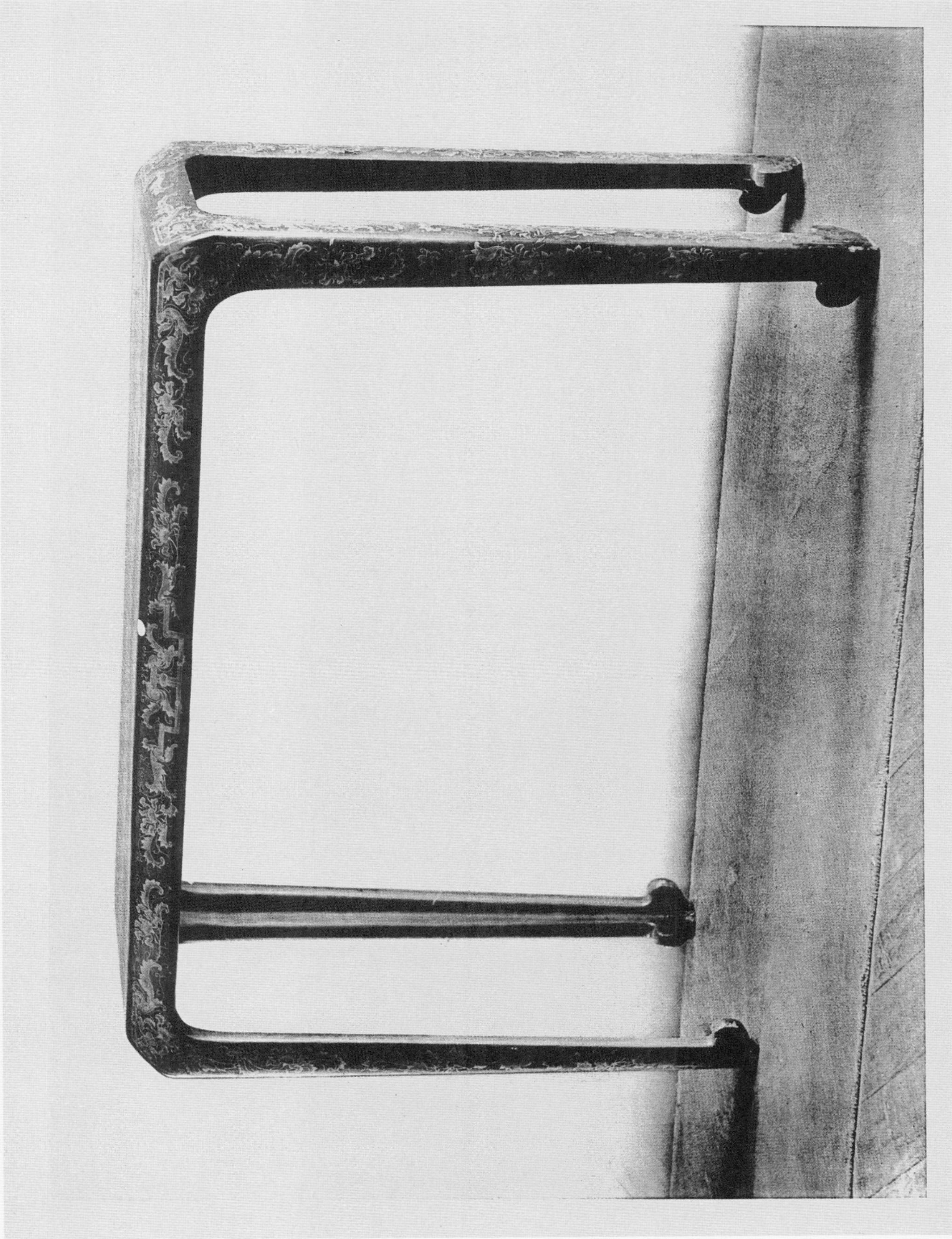

图版 37
PLATE XXXVII.

(1) 桌，浅黄漆，彩绘
    长 139.7 厘米　深 44.5 厘米　高 83.2 厘米

    保罗·马隆先生藏

(a) Table, buff lacquer, polychrome decoration. Length, 4 feet 7 inches; depth, 1 foot 5½ inches; height, 2 feet 8¾ inches.

In the possession of Paul Mallon, Esq.

(2) 上图桌的面
(b) Top of above table.

(3) 矮桌，黑漆，彩绘及嵌螺钿
    长 95.3 厘米　深 63.5 厘米　高 25.7 厘米

    保罗·马隆先生藏

(c) Low table, black lacquer, polychrome decoration and mother-of-pearl inlaid and carved. Length, 3 feet 1½ inches; depth, 2 feet 1 inch; height, 10⅛ inches.

In the possession of Paul Malon, Esq.

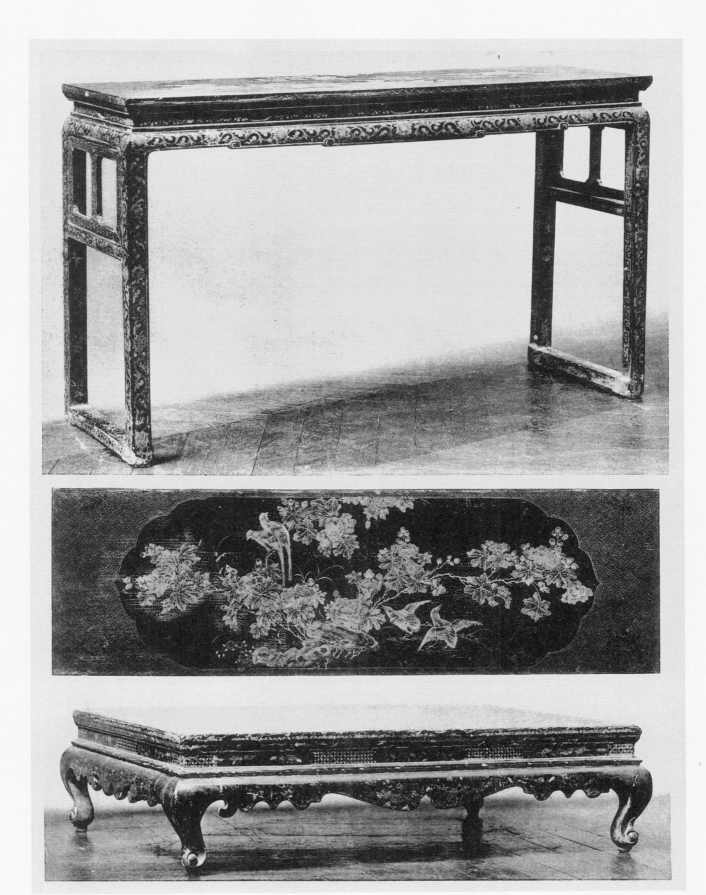

图版 38
PLATE XXXVIII.

罗汉床，红漆，彩绘
长 233.4 厘米　宽 139.7 厘米　高 109.8 厘米

卢吴公司藏

Bed of State, red lacquer, poly-chrome decoration. Length, 7 feet 7$\frac{7}{8}$ inches; breadth, 4 feet 7 inches; height, 3 feet 7$\frac{1}{4}$ inches.

In the possession of Messrs. Loo & Co.

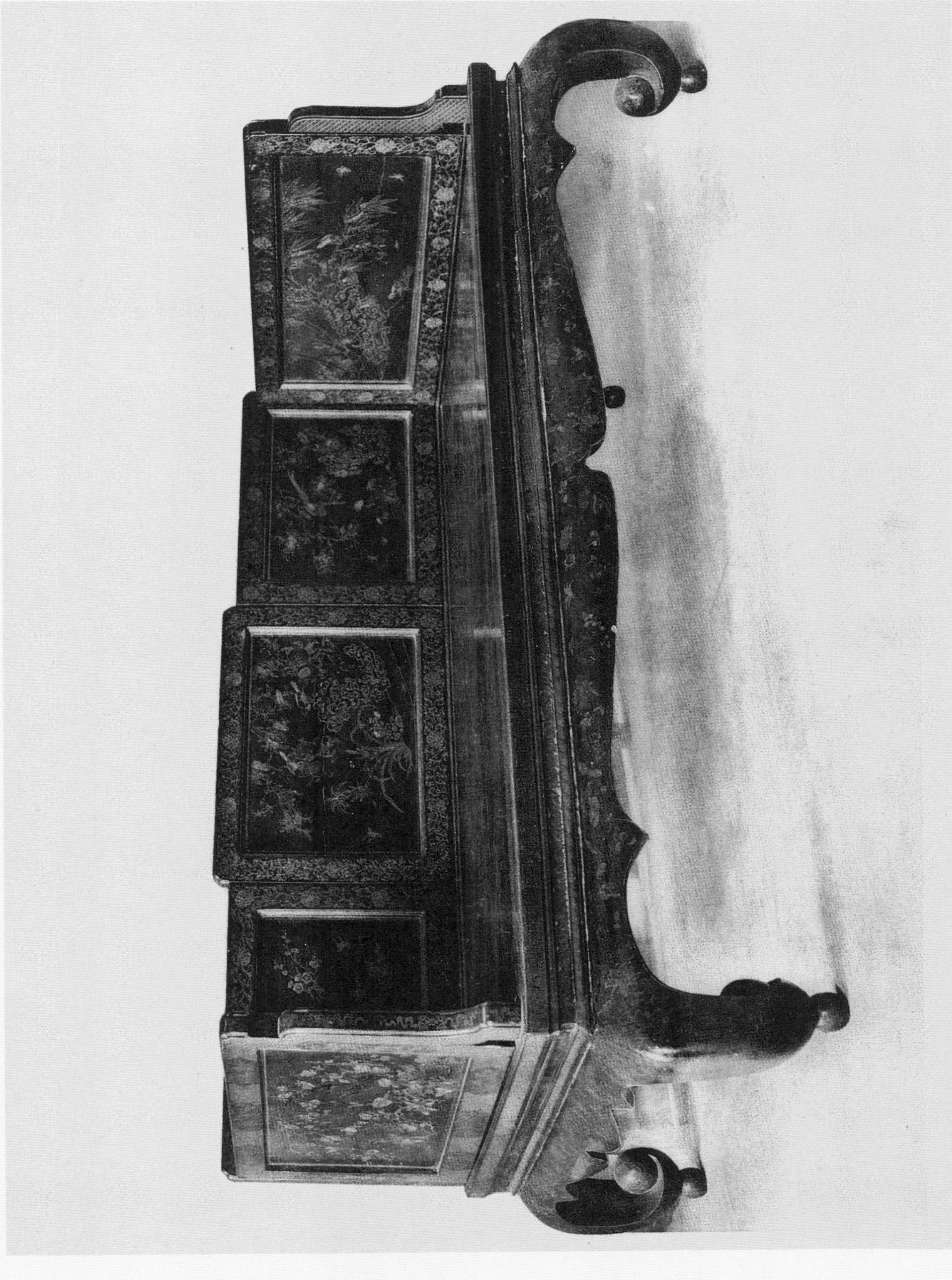

图版 39
PLATE XXXIX.

罗汉床，黑漆，描红、描金、嵌螺钿
长 223.5 厘米　宽 131.4 厘米　高 104.8 厘米

圣约翰·奥德利先生藏

Bed of State, black lacquer, red and gold decoration and mother-of-pearl inlaid and carved. Length, 7 feet 4 inches; breath, 4 feet $3\frac{3}{4}$ inches; height, 3 feet $5\frac{1}{4}$ inches.

In the possession of St. John Audley, Esq.

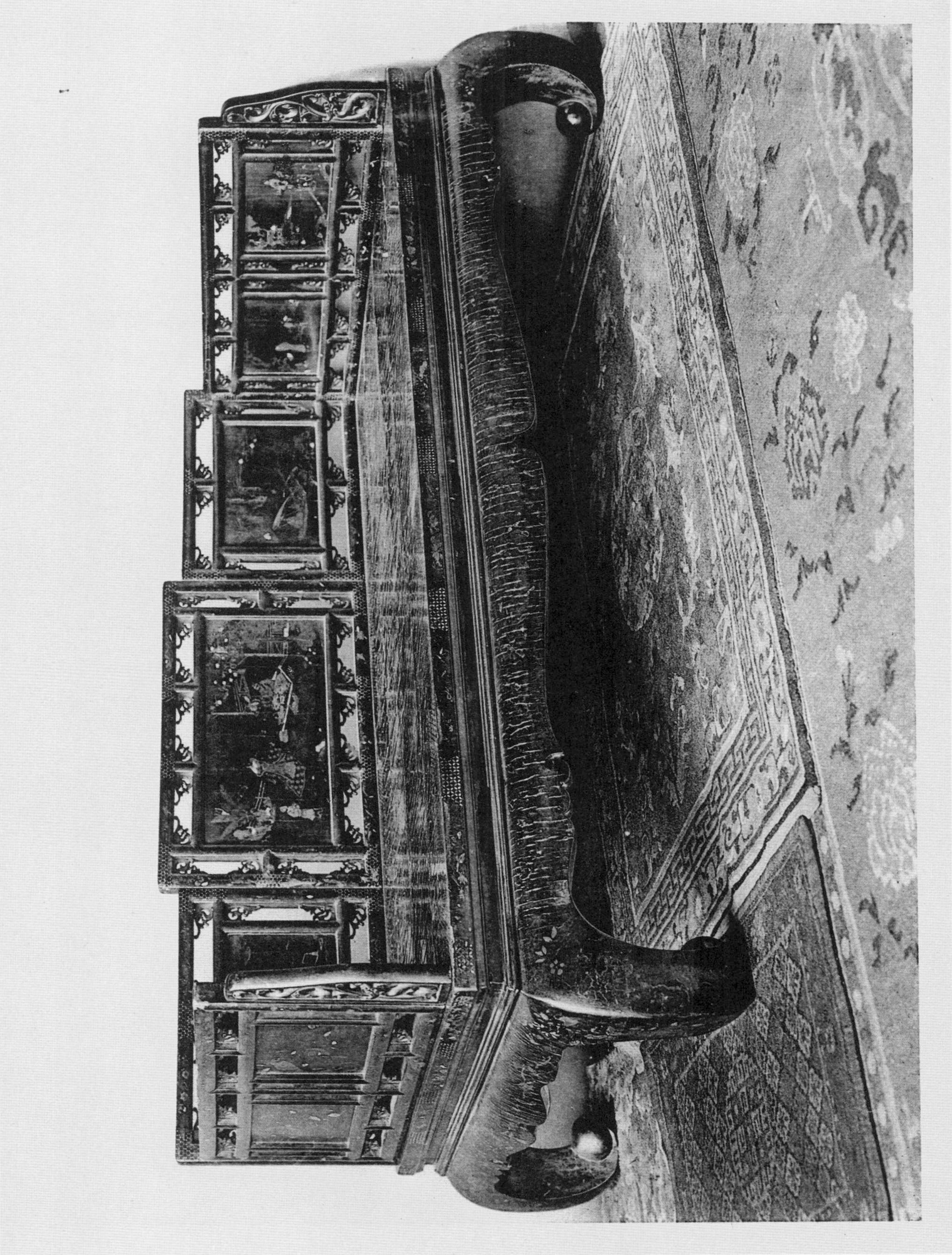

图版 40
PLATE XL.

大扶手椅，雕刻及黑漆
高 122.6 厘米　宽 125.7 厘米　深 78.1 厘米

圣约翰·奥德利先生藏

Large arm-chair, black lacquer and carved wood. Height, 4 feet $\frac{1}{4}$ inch; breadth, 4 feet $1\frac{1}{2}$ inches; depth, 2 feet $6\frac{3}{4}$ inches.

In the possession of St. John Audley, Esq.

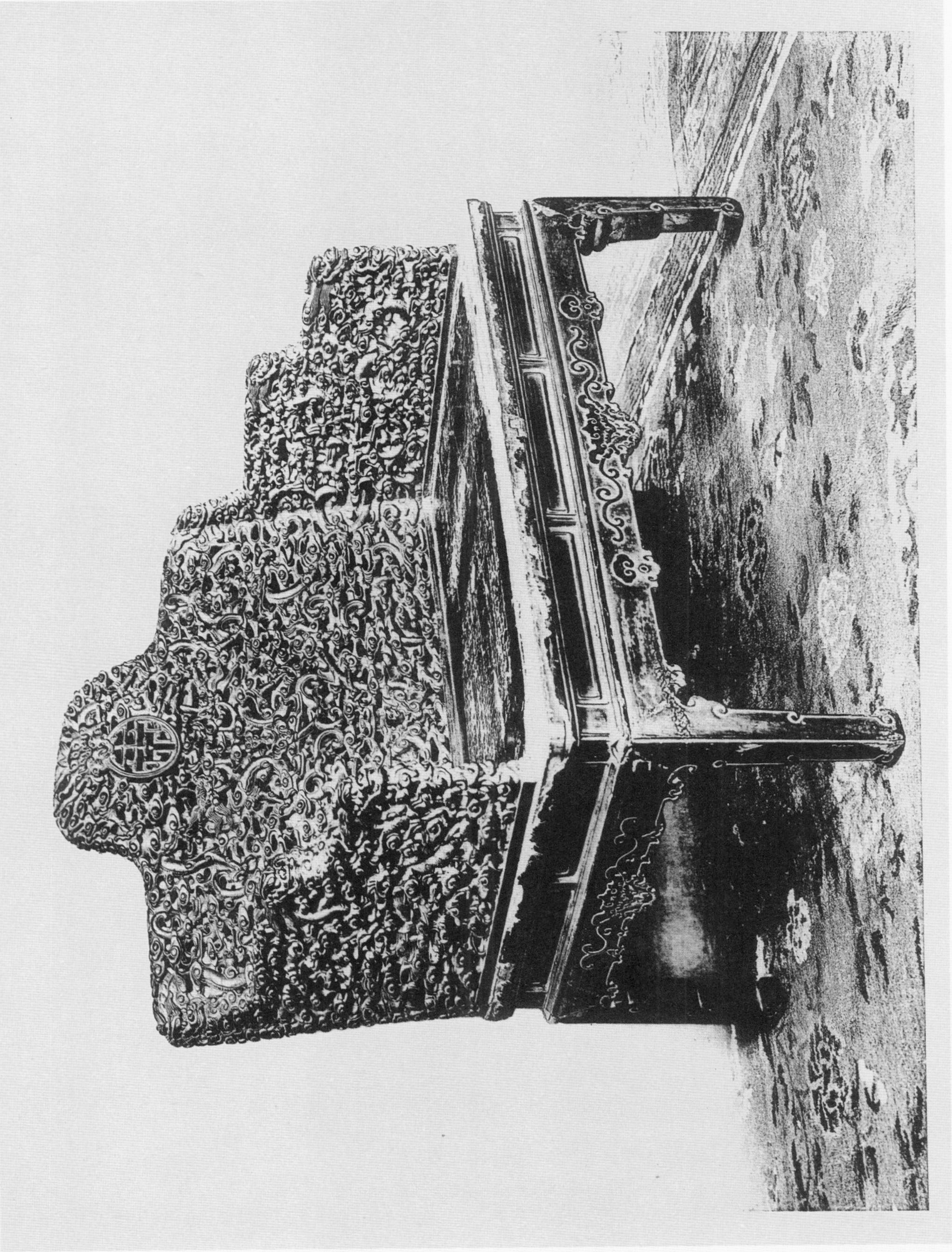

图版 41
PLATE XLI.

(1) 椅，红漆，彩绘
高 119.4 厘米　宽 57.5 厘米　深 43.8 厘米

沃尔希先生藏

(a) Chair, red lacquer, polychrome decoration. Height, 3 feet 11 inches; breadth, 1 foot $10\frac{5}{8}$ inches; depth, 1 foot $5\frac{1}{4}$ inches.

Worch Collecion.

(2) 扶手椅，红漆，彩绘
高 121.3 厘米　宽 69.2 厘米　深 54.6 厘米

沃尔希先生藏

(b) Arm-chair, red lacquer, polychrome decoration. Height, 3 feet $11\frac{3}{4}$ inches; breadth, 2 feet $3\frac{1}{4}$ inches; depth, 1 foot $9\frac{1}{2}$ inches.

Worch Collecion.

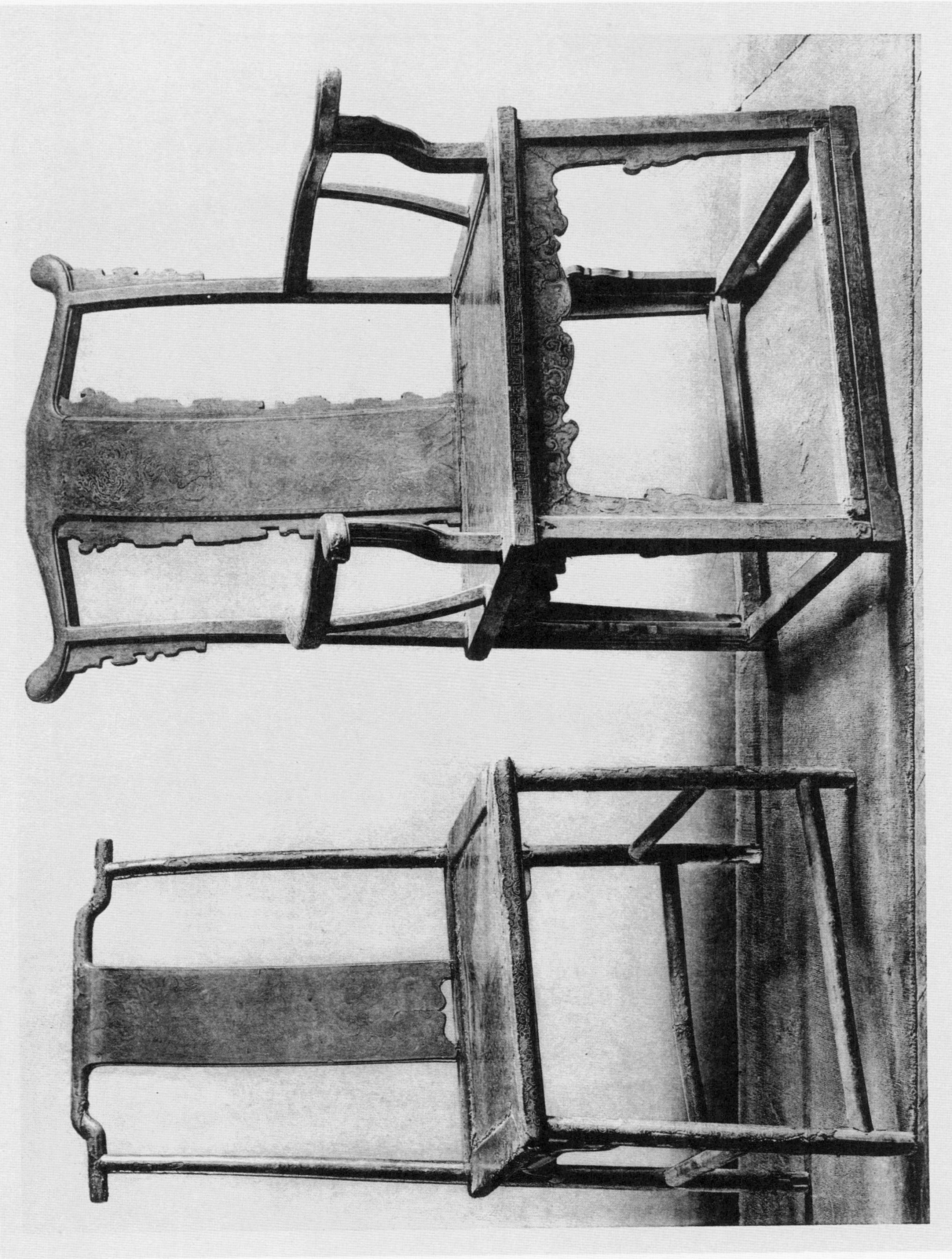

图版 42
PLATE XLII.

(1) 小扶手椅，红漆，描黑及描金
高 74.3 厘米　宽 49.5 厘米　深 36.8 厘米

沃尔希先生藏

(a) Small arm-chair, red lacquer, black and gold decoration. Height, 2 feet $5\frac{1}{4}$ inches; breadth, 1 foot $7\frac{1}{2}$; depth, 1 foot $2\frac{1}{2}$ inches.

Worch Collecion.

(2) 扶手椅，红漆，彩绘
高 95.9 厘米　宽 66.4 厘米　深 52.7 厘米

卢吴公司藏

(b) Arm-chair, red lacquer, polychrome decoration. Height, 3 feet $1\frac{3}{4}$ inches; breath, 2 feet $2\frac{1}{8}$ inches; depth, 1 foot $8\frac{3}{4}$ inches.

In the possession of Messrs. Loo & Co.

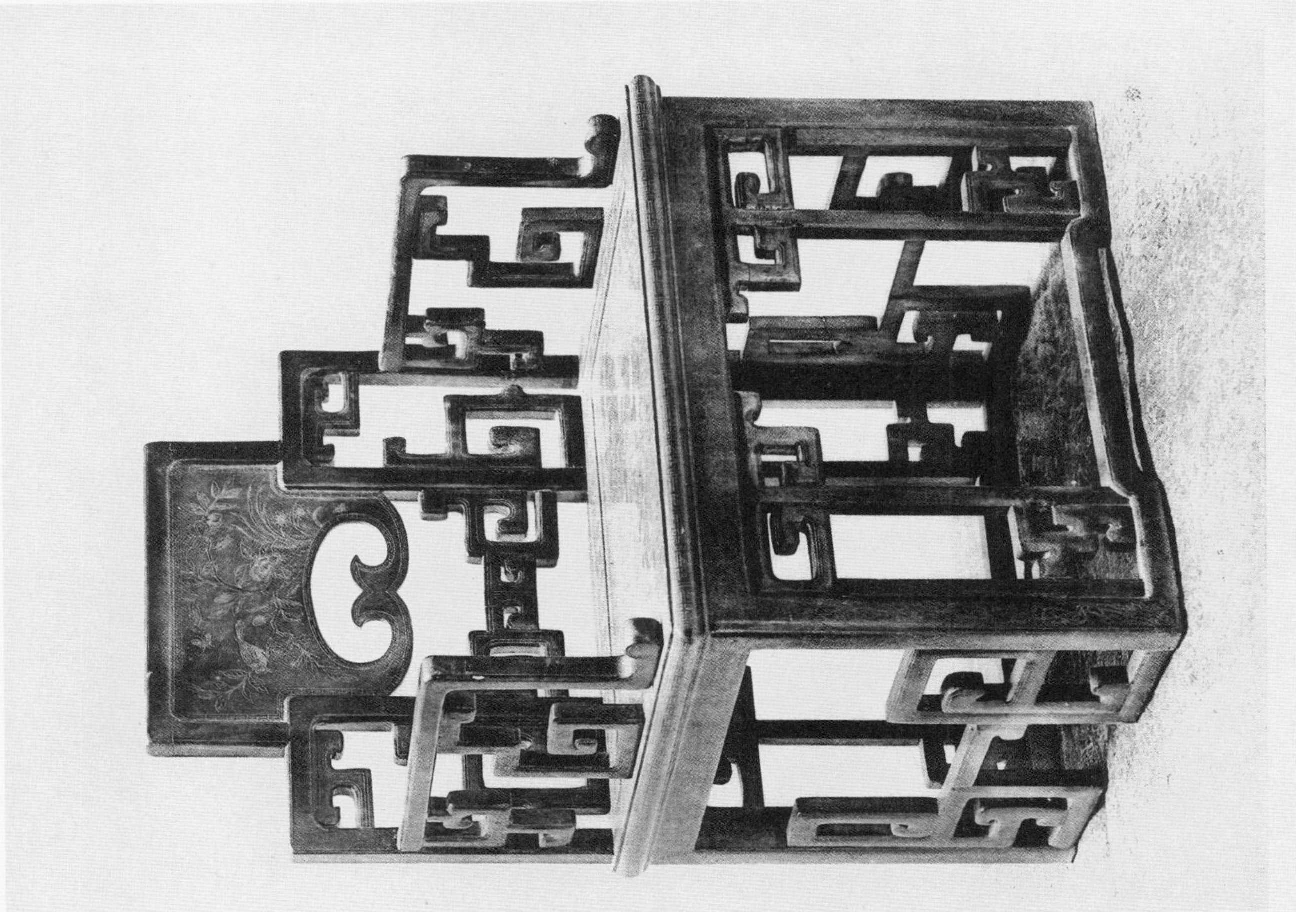
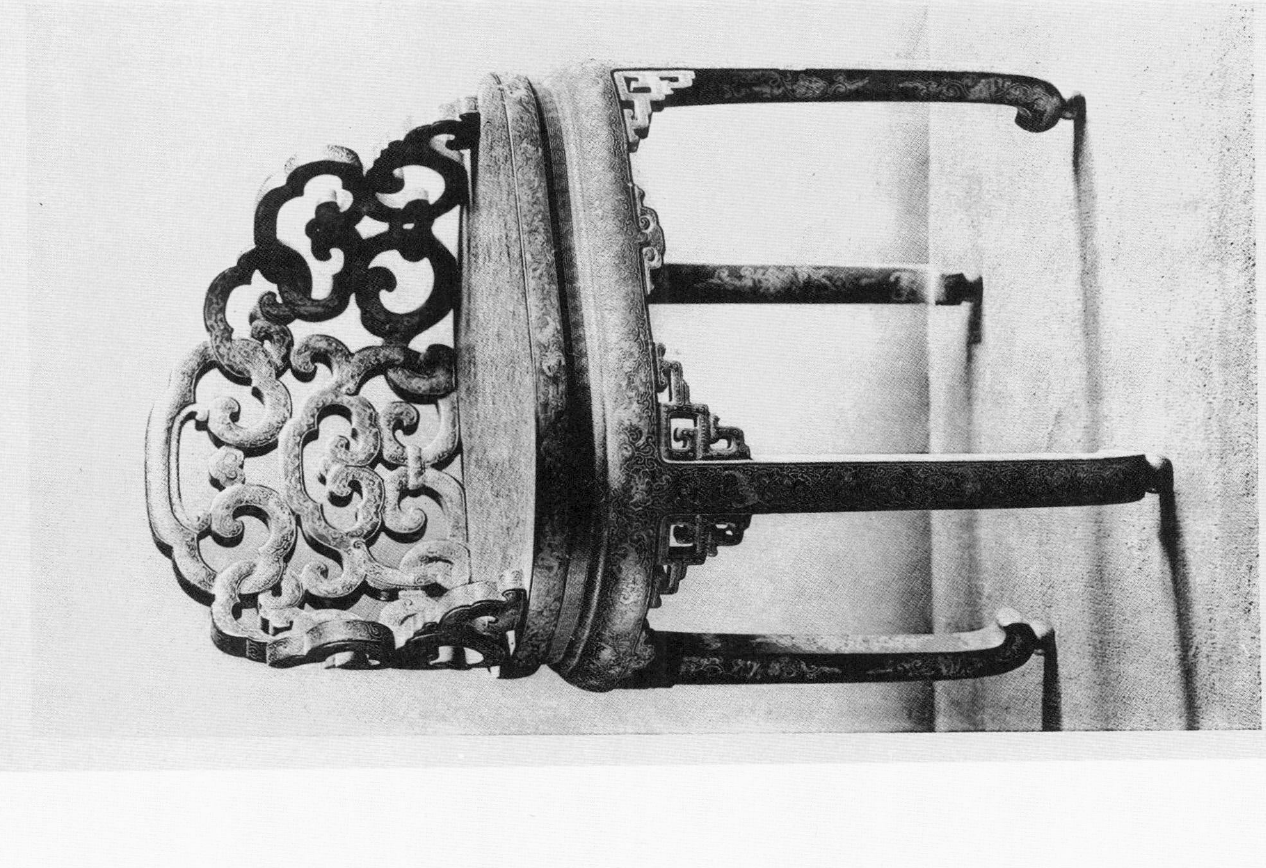

图版 43
PLATE XLIII.

(1) 扶手椅，黑漆，描金
    高 103.8 厘米　宽 82.2 厘米　深 56.5 厘米

    保罗·马隆先生藏

(a) Arm-chair, black lacquer, gold decoration. Height, 3 feet $4\frac{7}{8}$ inches; breadth, 2 feet $8\frac{3}{8}$ inches; depth, 1 foot $10\frac{1}{4}$ inches.

In the possession of Paul Mallon, Esq.

(2) 扶手椅，黑漆，描金
    高 102.9 厘米　宽 81.3 厘米　深 66.4 厘米

    保罗·马隆先生藏

(b) Arm-chair, black lacquer, gold decoration. Height, 3 feet $4\frac{1}{2}$ inches; breadth, 2 feet 8 inches; depth, 2 feet $2\frac{1}{8}$ inches.

In the possession of Paul Mallon, Esq.

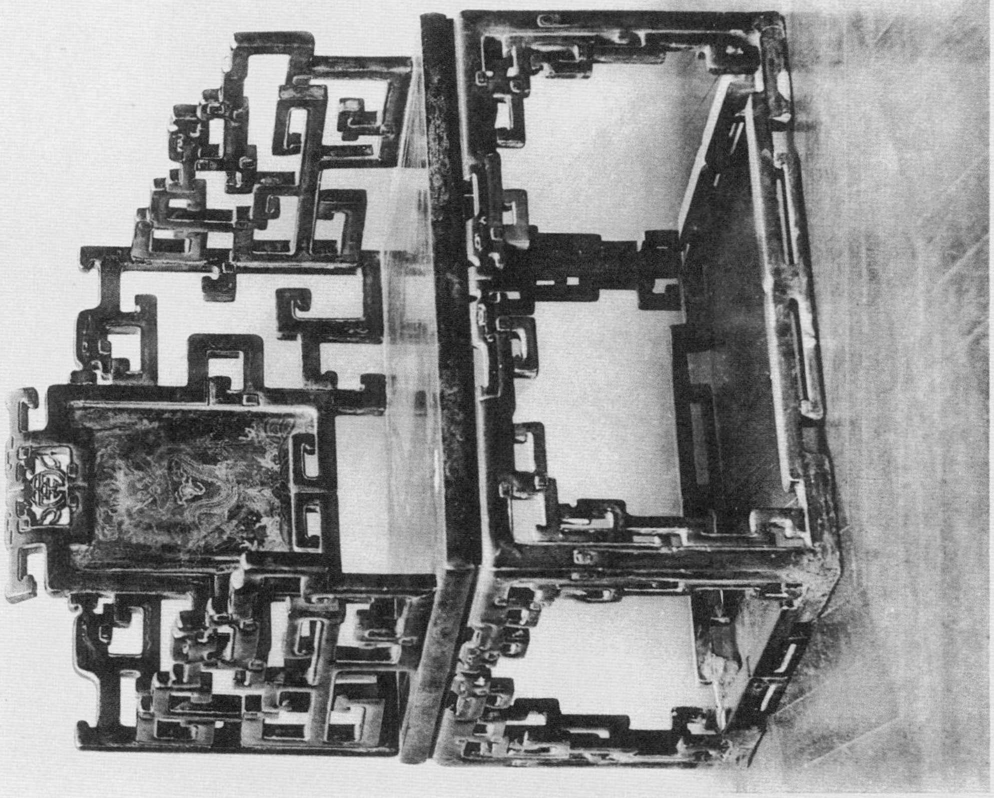
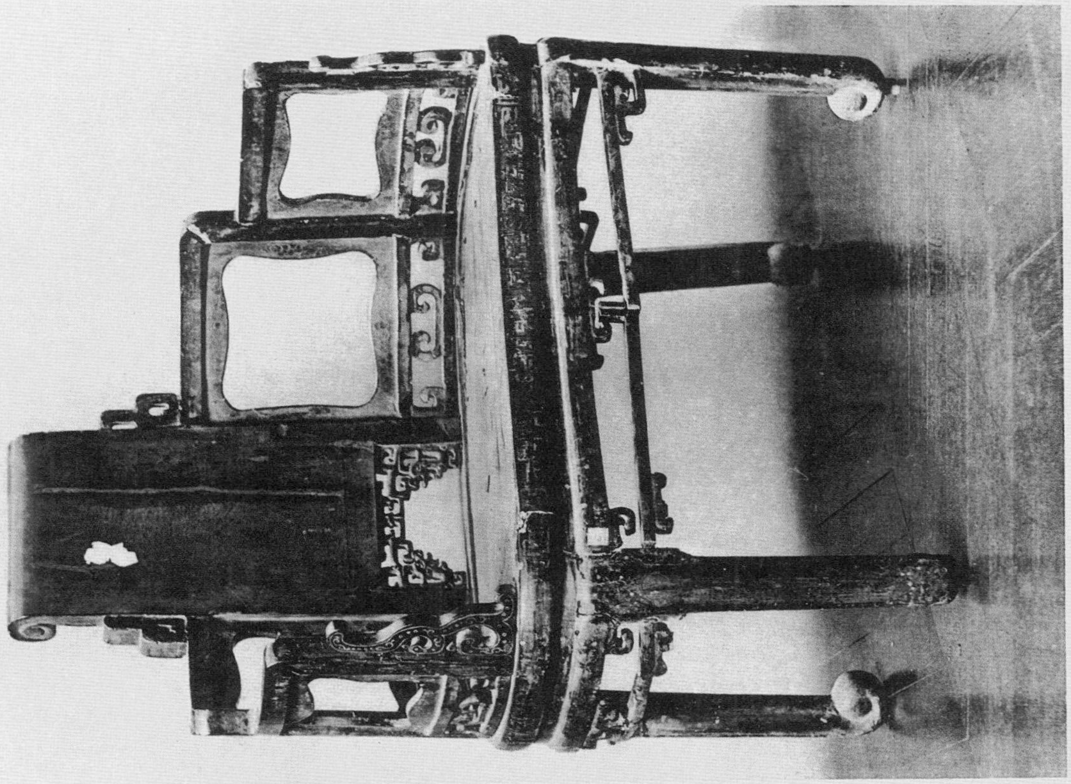

图版 44
PLATE XLIV.

(1) 扶手椅，黑漆，描金
高 92.1 厘米　宽 53.3 厘米　深 46.4 厘米

卢吴公司藏

(a) Arm-chairs, black lacquer, gold decoration. Height, 3 feet ¼ inch; breath, 1 foot 9 inches; depth, 1 foot 6 ¼ inches.

In the possession of Messrs. Loo & Co.

(2) 扶手椅，黑漆，描金
高 95.3 厘米　宽 59.4 厘米　深 45.7 厘米

中国通运公司藏

(b) Arm-chairs, black lacquer, gold decoration. Height, 3 feet 1½ inches; breadth, 1 foot 11⅛ inches; depth, 1 foot 6 inches.

In the possession of the Chinese Tongying Co.

图版 45
PLATE XLV.

(1) 扶手椅，黑漆，描金
高 99.7 厘米　宽 61.6 厘米　深 47.6 厘米

卢吴公司藏

(a) Arm-chair, black lacquer, gold decoration. Height, 3 feet $3\frac{1}{4}$ inches; breadth, 2 feet $\frac{1}{4}$ inch; depth, 1 foot $6\frac{3}{4}$ inches.

In the possession of Messrs. Loo & Co.

(2) 椅，黑漆，描金
高 103.8 厘米　宽 51.4 厘米　深 42.5 厘米

圣约翰·奥德利先生藏

(b) Chair, black lacquer, gold decoration. Height, 3 feet $4\frac{7}{8}$ inches; breadth, 1 foot $8\frac{1}{4}$ inches; depth, 1 foot $4\frac{3}{4}$ inches.

In the possession of St. John Audley, Esq.

图版 46
PLATE XLVI.

(1) 鼓凳，浅黄漆，描红及描金
高 46.4 厘米　径 43.8 厘米

卢吴公司藏

(a) Stool, buff lacquer, red and gold decoration. Height. 1 foot 6¼ inches; diameter, 1 foot 5¼ inches.

In the possession of Messrs. Loo & Co.

(2) 鼓凳，红漆，描黑及描金
高 45.7 厘米　径 36.8 厘米

沃尔希先生藏

(b) Stool, red lacquer, black and gold decoration. Height, 1 foot 6 inches; diameter, 1 foot 2½ inches.

Worch Collection.

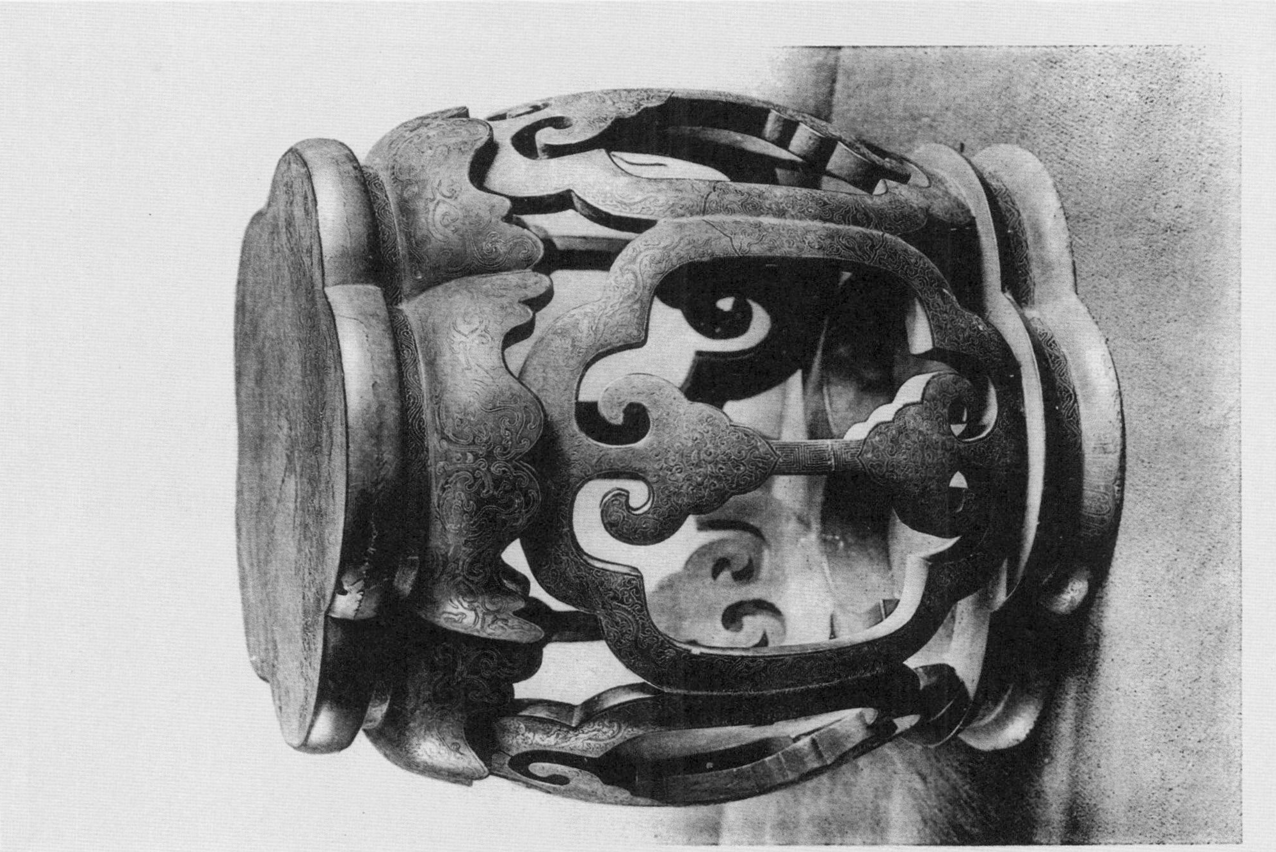
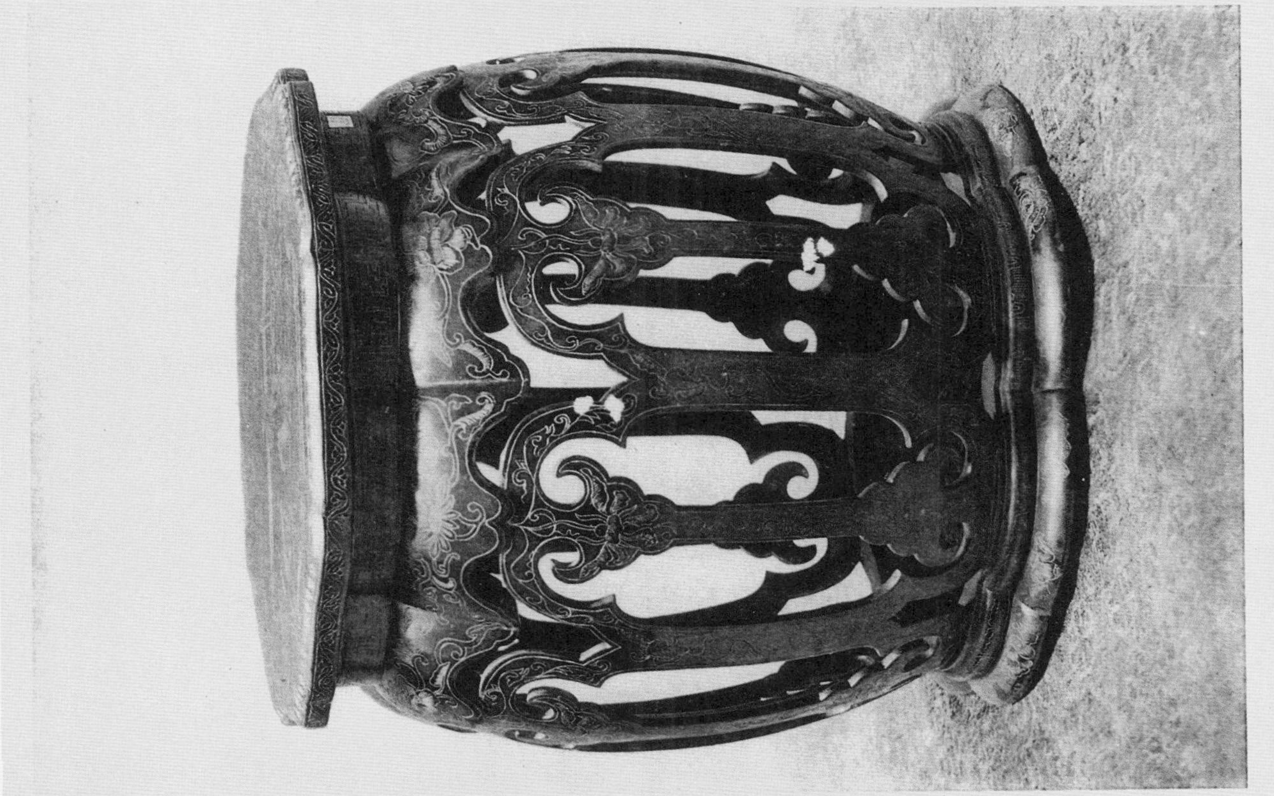

图版 47
PLATE XLVII.

(1) 鼓凳，黑漆，描金
高 49.5 厘米　上径 34.6 厘米

保罗·马隆先生藏

(a) Stool, black lacquer, gold decoration. Height, 1 foot 7½ inches; upper diameter, 1 foot 1⅝ inches.

In the possession of Paul Mallon, Esq.

(2) 灯架，红漆，描金
高 219.7 厘米

卢吴公司藏

(b) Lamp-stand, red lacquer, gold decoration. Height, 7 feet 2½ inches.

In the possession of Messrs. Loo & Co.

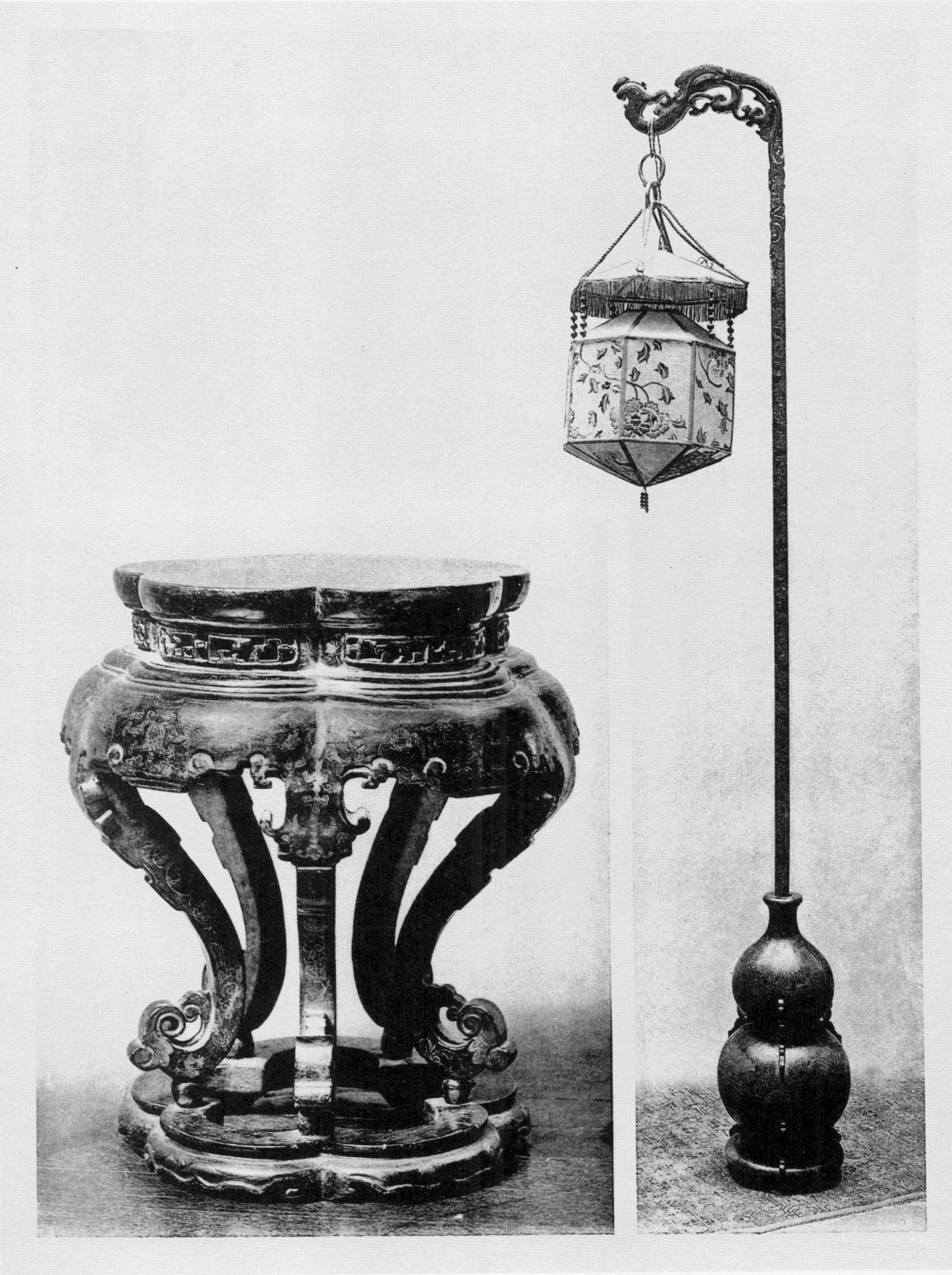

图版 48
PLATE XLVIII.

(1) 几，浅黄漆，描红及描黑
高 94 厘米　径 51.4 厘米

沃尔希先生藏

(a) Console, buff lacquer, red and black painted decoration. Height, 3 feet 1 inch; diameter, 1 foot 8¼ inches.

Worch Collection.

(2) 几，黑漆，描红及描金
高 88.3 厘米　宽 36.8 厘米

圣约翰·奥德利先生藏

(b) Console, black lacquer, red and gold decoration. Height, 2 feet 10¾ inches; breadth, 1 foot 2½ inches.

In the possession of St. John Audley, Esq.

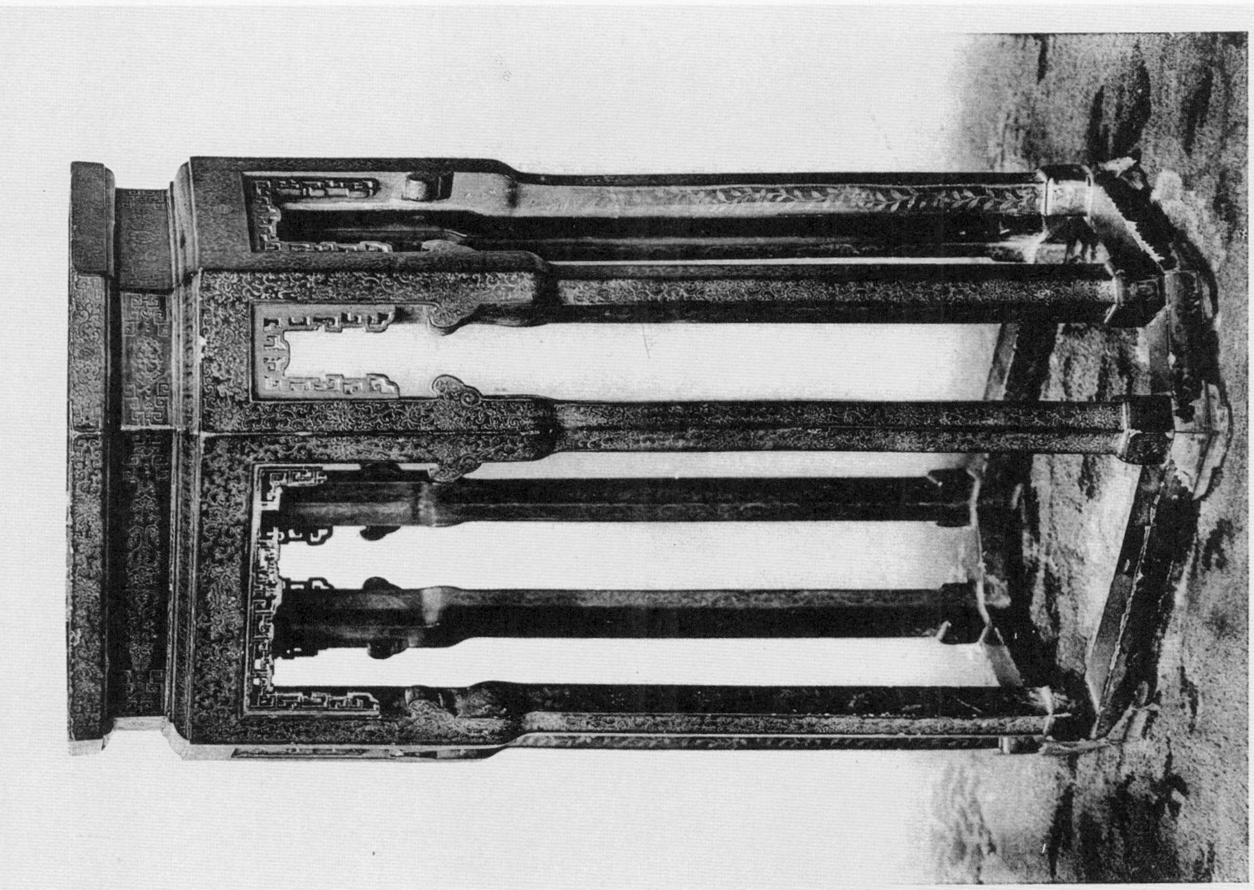
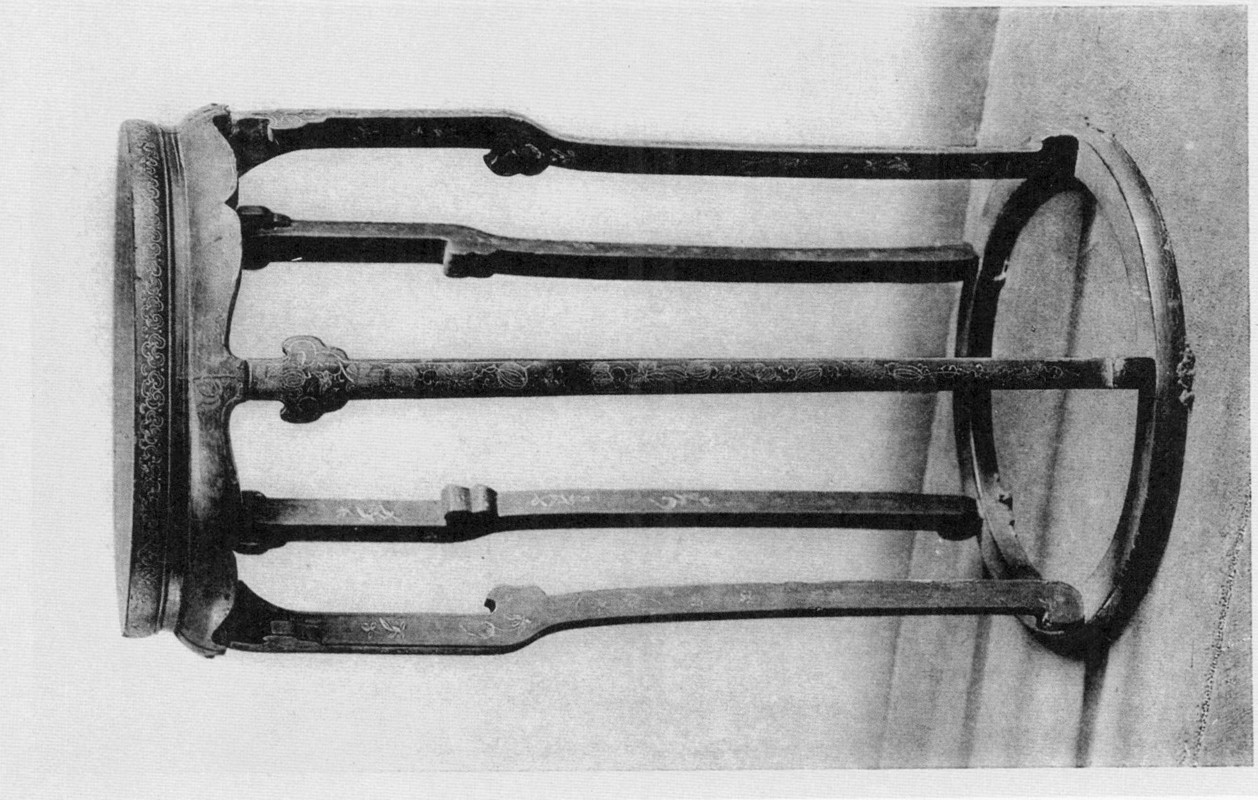

图版 49
PLATE XLIX.

12 扇大屏风，黑漆，彩绘
高 269.2 厘米　宽 599.4 厘米

奥迪隆·罗什先生藏

Large 12-leaf screen, black lacquer, polychrome decoration. Height, 8 feet 10 inches; breadth, 19 feet 8 inches.

In the possession of O. Roche, Esq.

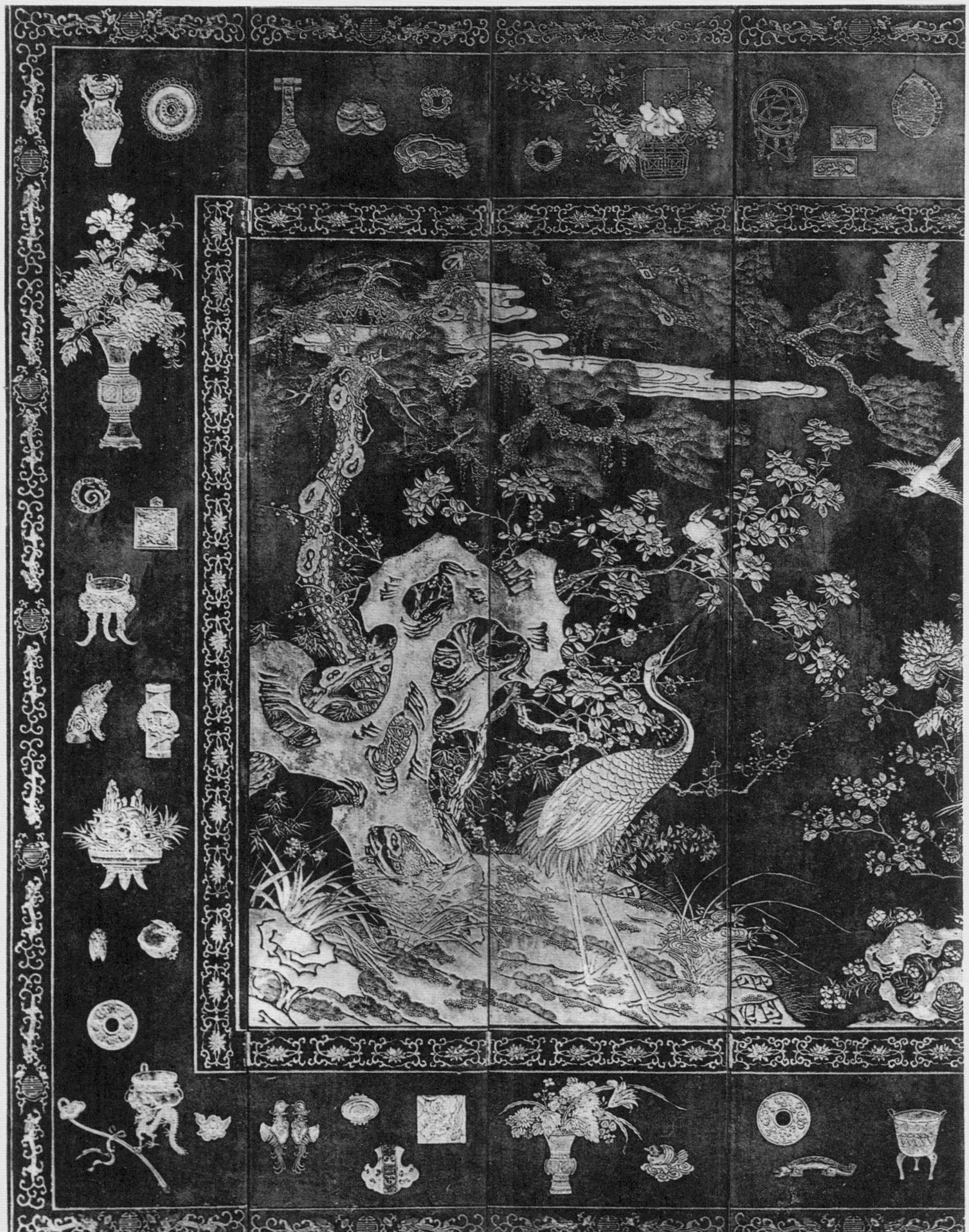

图版 50
PLATE L.

屏风的另一部分

Another part of the same screen.

图版 50

PLATE L

另外的另一部分

Another part of the same screen.

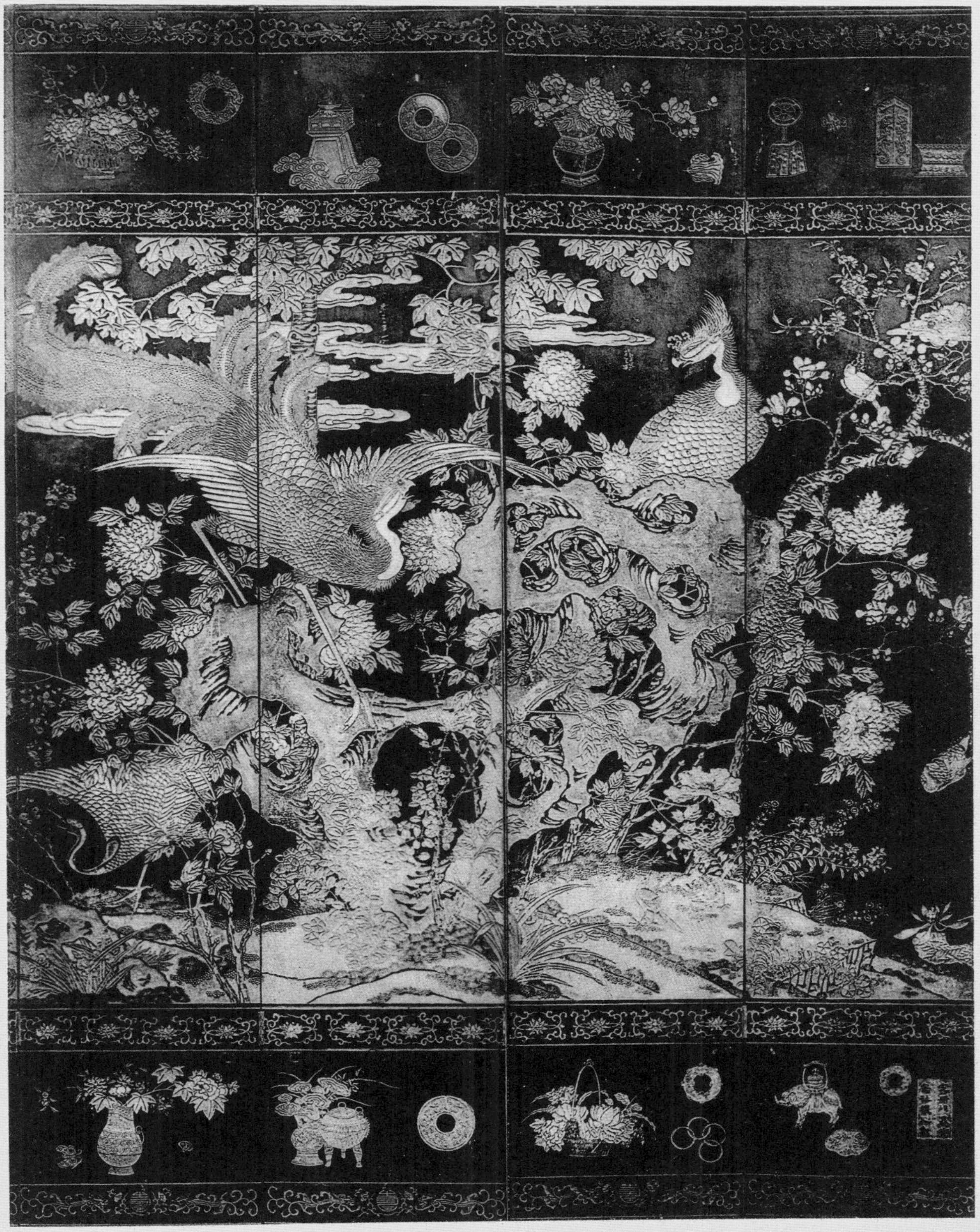

图版 51
PLATE LI.

屏风的另一部分

Another part of the same screen.

图版 51

PLATE LI

同一街的另一部分

Another part of the same street.

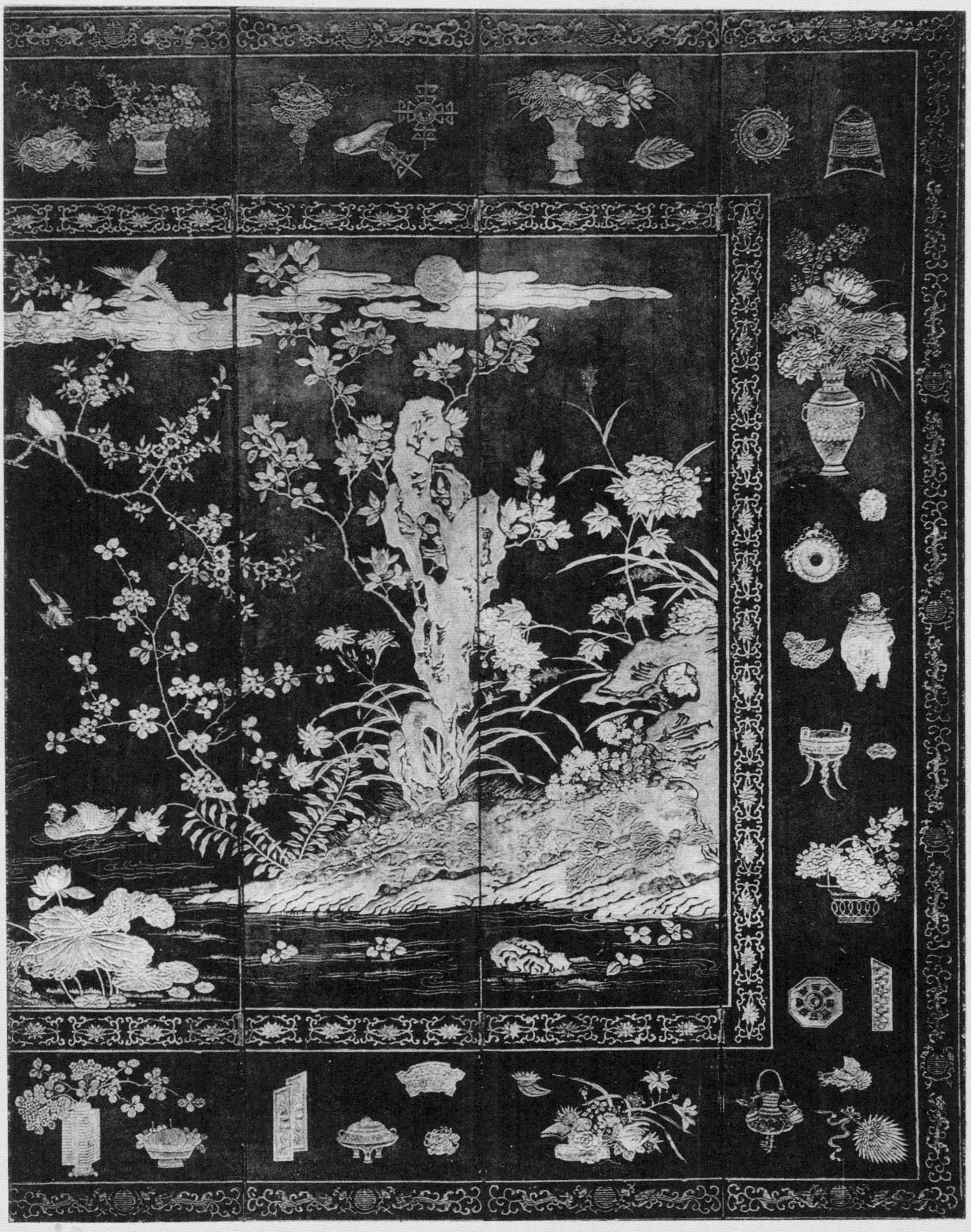

图版 52
PLATE LII.

12 扇大屏风，黑漆，彩绘
高 184.2 厘米　宽 574.7 厘米

中国通运公司藏

Large 12-leaf screen, black lacquer, polychrome decoration. Height, 6 feet ½ inch; breadth, 18 feet 10¼ inches.

In the possession of the Chinese Tongying Co.

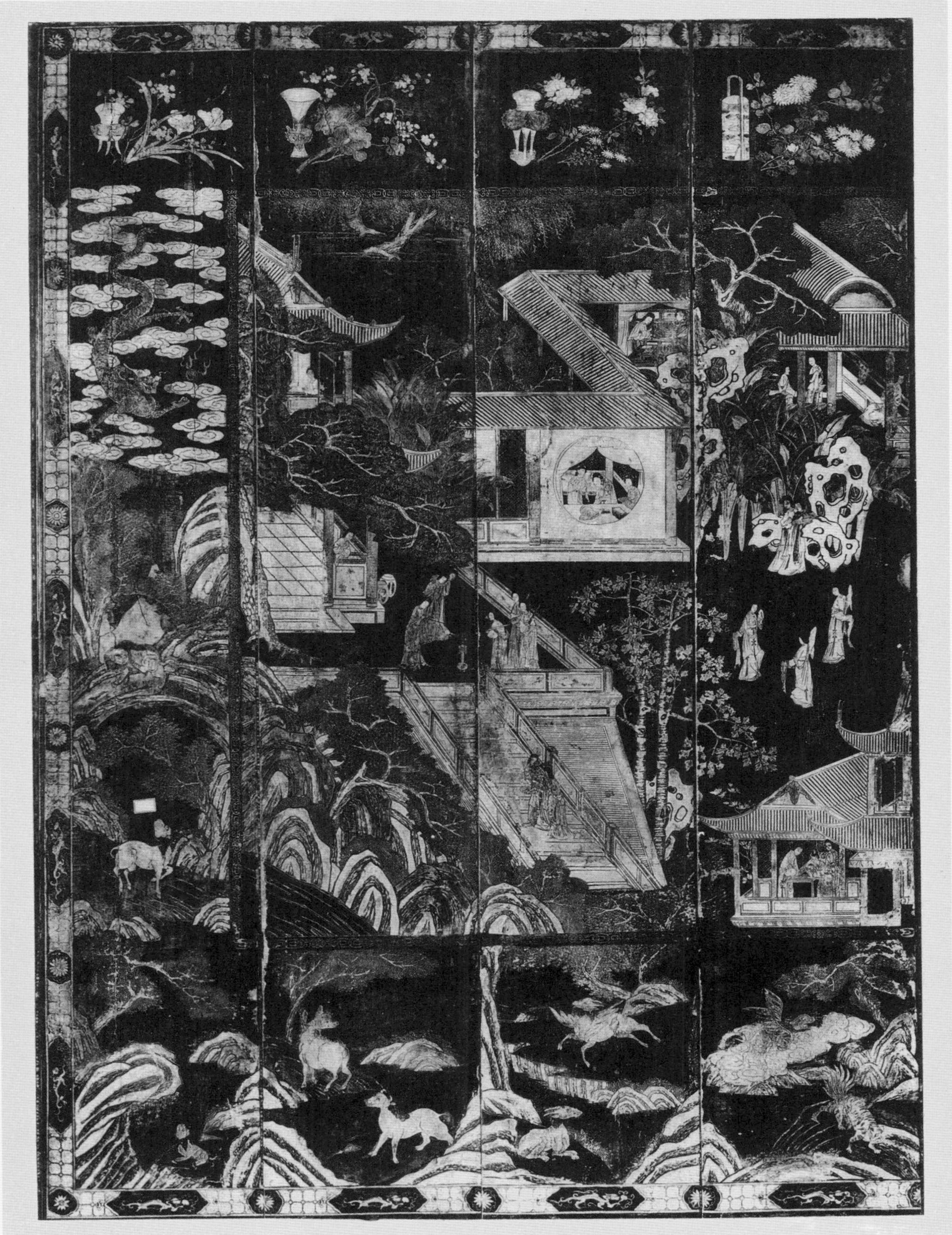

图版 53
PLATE LIII.

屏风的另一部分
Another part of the same screen.

图版 53
PLATE LIII

景观的另一部分
Another part of the same scene.

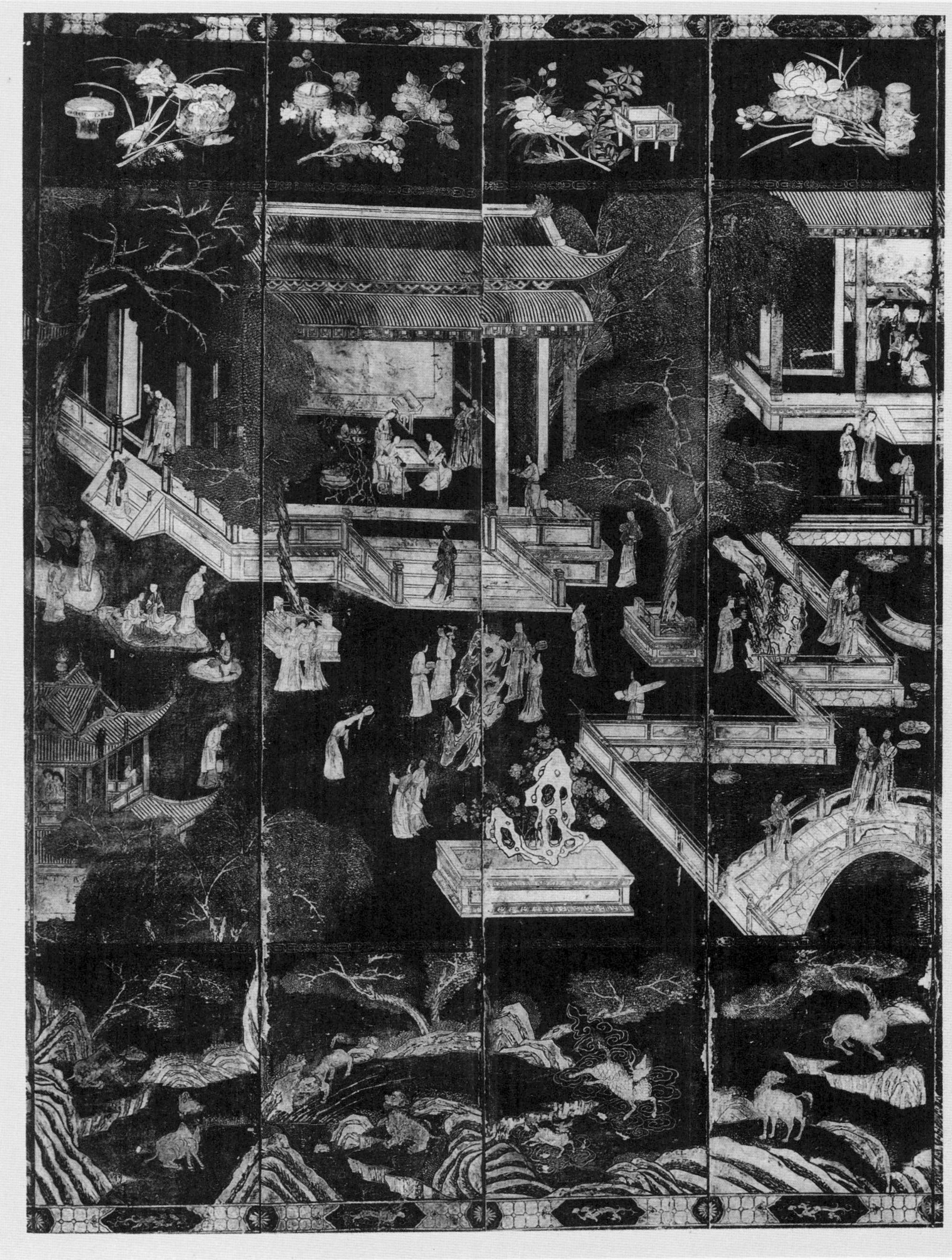

图版 54
PLATE LIV.

屏风的另一部分

Another part of the same screen.

圖版 54
PLATE LIV

圖版說明一部分
Another part of the same screen.

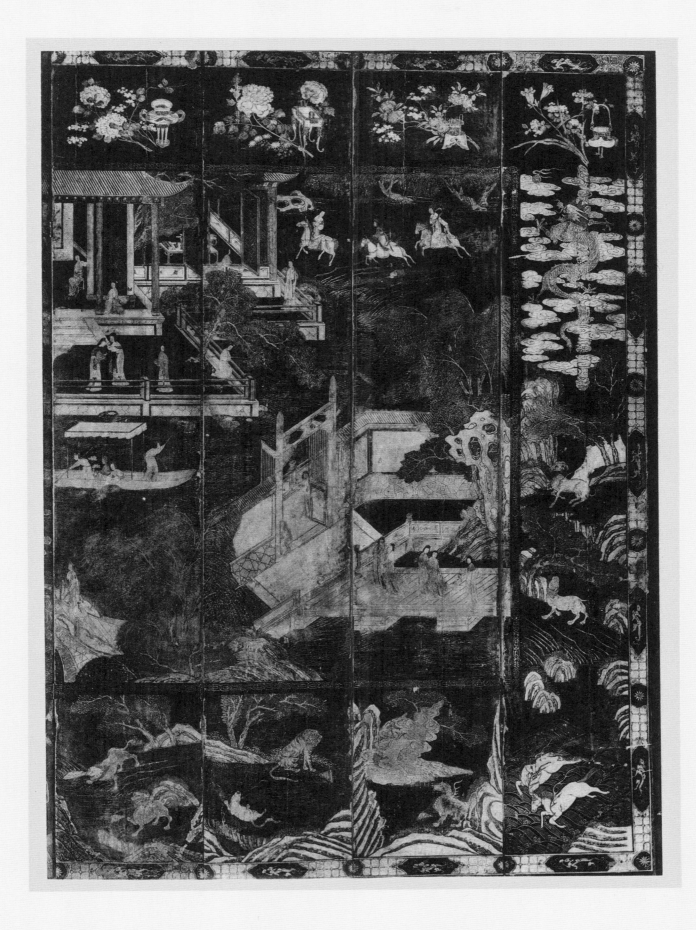

图书在版编目（CIP）数据

法国旧藏中国家具实例／（英）塞斯辛基编著．－北京：故宫出版社，2013.7
（中国家具经典图书辑丛）
ISBN 978-7-5134-0358-0

Ⅰ．①法… Ⅱ．①塞… Ⅲ．①家具－中国－古代－图集 Ⅳ．① TS666.202-64

中国版本图书馆 CIP 数据核字 (2012) 第 308780 号

责任编辑：张志辉
翻　　译：方　妍　张　萍
装帧设计：王　梓
出版发行：故宫出版社
　　　　　地址：北京东城区景山前街4号　邮编：100009
　　　　　电话：010-85007808　010-85007816　传真：010-65129479
　　　　　网址：www.culturefc.cn
　　　　　邮箱：ggcb@culturefc.cn
制版印刷：北京圣彩虹制版印刷技术有限公司
开　　本：635×965毫米　1/8
印　　张：31.5
版　　次：2013年7月第1版
　　　　　2013年7月第1次印刷
书　　号：978-7-5134-0358-0
定　　价：460.00元